Donald McIntosh

Diseases of Swine

Written as a text book for the veterinary surgeon, student and swine grower

Donald McIntosh

Diseases of Swine

Written as a text book for the veterinary surgeon, student and swine grower

ISBN/EAN: 9783337328726

Printed in Europe, USA, Canada, Australia, Japan

Cover: Foto ©berggeist007 / pixelio.de

More available books at **www.hansebooks.com**

DISEASES OF SWINE.

WRITTEN AS A TEXT BOOK

FOR THE

Veterinary Surgeon, Student and Swine Grower.

D. McINTOSH, V. S.,

PROFESSOR OF VETERINARY SCIENCE IN THE UNIVERSITY OF ILLINOIS.
AUTHOR OF "DISEASES OF HORSES AND CATTLE."

PREFACE.

DISEASES of the pig have not been hitherto adequately represented in veterinary literature as they should have been, especially in this country where the growing of pigs is one of the principal industries, represented by hundreds of thousands of dollars. I have been surprised that some of my professional brethren have not taken up this work, but I suppose that lack of time and opportunity prevented them from so doing. Some ten years ago I was urged by some of my friends to undertake such work, but having so little literature on the subject to cull from, I hesitated to do so, but since coming to Illinois I have had ample opportunity of investigating all the diseases of swine.

The subjects dealt with are based on science and confirmed by experience, so that the reader will not lose time in reading theories which are not confirmed by facts. There are a few diseases of swine, such as thumps, partial paralysis of the hind quarters and canker of the mouth, which will in the majority of cases prove fatal. These diseases have been thoroughly investigated and the proper remedies discovered, and if administered as set forth in this treatise the majority of the affected pigs will recover. I have, with the assistance of Doctor Burrill, Professor of Bacteriology in the University of Illinois, investigated extensively "hog cholera," and have made considerable headway in treating the disease successfully. These results and treatment will be found in this volume. It also contains as an introduction an article on the health of the pig which will be of great service to the swine breeder. The book contains a number of illustrations which will be helpful to the reader. My hope is that this manual will fill a long felt want to the veterinary surgeon and swine grower.

INDEX.

	PAGE.
Acne	152
Amaurosis	158
Anemia	132
Angina, gangrenous	141
Ani, prolapsus	63
Anthrax, neck	139
Apoplexy	130
Appetite, morbid	43
Aphtha, sporadic	27
Aphthous fever	137
Arteries, diseases of	129
Ascaris Suilla	84
Atrophy	126
Atrophy of the Kidney	196
Atrophy of the Liver	72
Black Teeth	31
Bladder, inflammation of	198
Bleeding Fungus	207
Blood with the Urine	192
Bowels, inflammation of	47
Bowels, obstructions of	59
Bowels, twisting of	60
Brain, inflammation of	176
Brain, inflammation of the membrane of	176
Bronchitis	102
Bruises	206
Calculi, urinary	202
Canina, rabies	185
Canker of the Nose and Face	144
Catarrh, malignant	97
Chorea	179
Chronic Hepatitis	70
Chronic Inflammation of Nostrils	96
Chronic Cough	120
Chronic Laryngitis	101
Chronic Rheumatism	173
Colic	56
Congestion of the Lungs	106
Conjunctivitis	157
Constipation	57
Coryza	94
Cough, nervous	102
Cysts-Hyatids	73
Cystitis	198
Diarrhea	53
Difficult Parturition	208
Diseases of Arteries and Veins	129
Diseases of Eyeball	157
Diseases of Generative Organs	208
Diseases of Heart	123
Diseases of Intestines	47
Diseases of Liver of the Pig	69
Diseases of Nervous System	176
Diseases of Organs of Mastication	26
Diseases of Respiratory Organs	94

	PAGE.
Diseases of Stomach of the Pig	34
Diseases of Urinary Organs	189
Dysentery	49
Emphysema	117
Enteritis	47
Enuresis	201
Erysipelas	166
Erysipelas, gangrenous	140
Eustrongylus Gigas	89
Eyeball, diseases of	157
Face, canker of	144
Fatty Liver	72
Feet, sore	155
Fever, scarlet	160
Flesh, proud	205
Functional Diseases of the Heart	127
Fungous Growths	207
Fungus, bleeding	207
Gangrenous Angina	141
Gangrenous Erysipelas	140
Gangrenous Inflammation of the Mouth	30
Gastritis	36
General Observations	78
Generative Organs, diseases of	208
Gravel	202
Growths, fungous	207
Heart, diseases of	123
Heart, hypertrophy of	126
Hematuria	192
Hematuria, idiopathic	193
Hematemesis	44
Hemorrhoids	62
Hemoptyses	121
Hepatitis	69
Hepatitis, chronic	70
Hernia, scrotal	76
Hernia, umbilical	75
Hog Cholera	221
Hydrophobia	185
Hypertrophy of Liver	71
Hypertrophy of Heart	126
Hypertrophy of Kidney	196
Hydrocele	219
Idiopathic Hematuria	193
Idiopathic Tetanus	181
Incontinence of Urine	201
Induration	72
Inflammation of the Brain	176
Inflammation of Membrane of the Brain	176
Inflammation of Bowels	47
Inflammation of Bladder	198
Inflammation of Kidney	190
Inflammation of Mouth	30
Inflammation of Nostrils	94
Inflammation of Testicles	218

	PAGE
Inflammation of Tongue	29
Inflammation of Udder	212
Infections of the Blood, purulent	134
Injuries	206
Internal Parasites	78
Intestines, diseases of	47
Ischuria	195
Jaundice	73
Kidney, atrophy of	196
Kidney, hypertrophy of	196
Kidney, inflammation of	190
Kidney, rupture of	197
Kidney Worm	89
Lacerated Wounds	204
Laryngitis	99
Laryngitis, chronic	101
Lice	147
Lichen	149
Liver of the pig	68
Liver, diseases of	69
Liver, fatty	72
Liver, waxy	72
Lungs, congestion of	106
Malignant Catarrh	97
Mammitis	216
Mange	145
Mastication, organs of, diseases of	26
Measles	163
Meat Brine	38
Morbid Appetite	43
Mastication, organs of	21
Mouth, gangrenous inflammation of	30
Neck Anthrax	139
Nephritis	190
Nervous Cough	102
Nervous Diseases of the Heart	127
Nose, canker of	144
Observations, general	78
Obstructions of the Bowels	59
Ophthalmia	157
Optic Nerve, paralysis of	158
Orchitis	218
Organs of Mastication	21
Organs of Mastication, diseases of	26
Oxurus Vernicularis	82
Palpitation	127
Paralysis	181
Paralysis, partial	183
Parasites, internal	78
Parturition, difficult	208
Pemphigus	151
Pericarditis	123
Peritonitis	64
Pharyngitis	99
Phrenitis	176
Piles	62
Plethora	131
Pleurisy	114
Pneumonia	109
Prolapsus Ani	63
Proud Flesh	205
Punctured Wounds	206
Purulent Infections of the Blood	134
Quinsy	98
Rabies	185
Rectum, stricture of the	61
Respiratory Organs, diseases of	94

	PAGE
Retention, vesical	200
Rheumatism	170
Rheumatism, chronic	173
Ring Worm	154
Round Worm	84
Rupture	75
Rupture of Kidney	197
Rupia	152
Scaly Diseases of the Skin	153
Scarlet Fever	160
Sclerostonum Dentatum Diesing	86
Scrotal Hernia	76
Scrotum, water in the	219
Serous Cysts-Hyatids	73
Simple Ulcers of the Stomach	40
Skin, diseases of	143
Sore Feet	155
Sore Teats	217
Sore Throat	99
Spiroptera Strongylina Rud	86
Sporadic Aphtha	27
Sprains	174
Sterility	219
Stomach of the Pig	32
Stomach of the Pig, diseases of	34
Stricture of Rectum	61
Strongylus Dentatum Rud	87
Strongylus Elognatus	88
Suppression of Urine	195
Swine Plague	221
Tape worm	77
Teats, sore	217
Teeth, black	31
Tetanus	181
Tetanus, idiopathic	181
Testicles, inflammation of	218
Thorn Headed Worm	80
Thread Worm	83
Throat, sore	99
Thumps	127
Tongue, inflammation of	29
Trecocephalus Dispar	83
Trichina Spiralis	90
Twisting of the Bowels	60
Ulcers	207
Ulcers of the Stomach, simple	40
Umbilical Hernia	75
Unhealthy Wounds	205
Urine, blood with	192
Urine, incontinence of	201
Urine, suppression of	195
Urinary Calculi	202
Urinary Organs, diseases of	189
Urticaria	148
Uterus, inversion of	215
Veins, diseases of	129
Vesical Retention	200
Vomiting	42
Warts	148
Water in the Scrotum	219
Waxy Liver	72
Worm, kidney	89
Worm, long thread	83
Worm, pin	82
Worm, ring	154
Worm, round	84
Worm, thorn headed	80
Wounds, lacerated	204
Wounds, punctured	206
Wounds, unhealthy	205

DISEASES OF SWINE.

CHAPTER 1.
The Organs of Mastication.

CHAPTER 2.
Diseases of the Organs of Mastication.

Sporadic Aphtha, Inflammation of the Tongue, Gangrenous Inflammation of the Mouth, Black Teeth.

CHAPTER 3.
Stomach of the Pig.

CHAPTER 4.
Diseases of the Stomach of the Pig.

Gastritis, Meat Brine, Simple Ulcers of the Stomach, Vomiting, Morbid Appetite, Hematemesis.

CHAPTER 5.
Diseases of the Intestines.

Enteritis, Inflammation of the Bowels, Dysentery, Diarrhea, Colic, Constipation, Obstructions of the Bowels, Twisting of the Bowels, Stricture of the Rectum, Hemorrhoids or Piles, Prolapsus Ani, Peritonitis.

CHAPTER 6.
Diseases of the Liver of the Pig.

Hepatitis, Chronic Hepatitis, Hypertrophy, Atrophy, Induration, Fatty Liver, Waxy Liver, Serous Cysts-Hyatids, Jaundice.

CHAPTER 7.
Hernia, Rupture.

Umbilical Hernia, Scrotal Hernia.

CHAPTER 8.
Internal Parasites of the Pig.

General Observations—Oxyrus Vernicularis (Pin Worm), Thorn Headed Worm, Trecocephalus Dispar (Long Thread Worm), Ascaris Suilla (Round Worm) Spiroptera Strongylina Rud, Strongylus Dentatum (Rud), Sclerostonum Dentatum Diesing, Strongylus Elognatus, Kidney Worm Eustrongylus Gigas, Trichina Spiralis.

CHAPTER 9.
Diseases of the Respiratory Organs.

Inflammation of the Nostrils or Coryza, Chronic Inflammation of the Nostrils or Ozena, Malignant Catarrh, Quinsy, Laryngitis, Pharyngitis (Sore Throat), Chronic Laryngitis, Nervous Cough, Bronchitis, Congestion of the Lungs, Pneumonia, Pleurisy, Emphysema, Chronic Cough, Hemoptysis.

CHAPTER 10.
Diseases of the Heart.

Pericarditis, Hypertrophy, Atrophy, Functional or Nervous Diseases of the Heart, Palpitation or Thumps,

CHAPTER 11.
DISEASES OF THE ARTERIES AND VEINS.

CHAPTER 12.
Apoplexy, Plethora, Anemia.

CHAPTER 13.
PURULENT INFLECTIONS OF THE BLOOD.

CHAPTER 14.
ANTHRAX OF THE PIG.
Neck Anthrax, Gangrenous Erysipelas, Gangrenous Angina.

CHAPTER 15.
DISEASES OF THE SKIN.
Canker of the Nose and Face, Mange, Urticaria, Lichen, Prurigo, Pemphigus, Rupia, Acne, Scaly Diseases of the Skin, Ring Worm, Lice, Warts, Sore Feet.

CHAPTER 16.
DISEASES OF THE EYEBALL.
Conjunctivitis or Simple Ophthalmia, Amaurosis or Paralysis of the Optic Nerve.

CHAPTER 17.
FEVERS.
Scarlet Fever, Measles.

CHAPTER 18.
ERYSIPELAS.

CHAPTER 19.
RHEUMATISM.
Chronic Rheumatism, Sprains.

CHAPTER 20.
DISEASES OF THE NERVOUS SYSTEM.
Phrenitis (Inflammation of the Brain), Meningitis (Inflammation of the Membrane of the Brain), Chorea, Tetanus, Idiopathic Tetanus, Paralysis, Partial Paralysis, Hydrophobia, Rabies, Rabies Canina.

CHAPTER 21.
DISEASES OF THE URINARY ORGANS.
Nephritis (Inflammation of the Kidneys), Hematuria (Blood with the Urine), Idiopathic Hematuria, Ischuria (Suppression of Urine), Atrophy of the Kidneys, Hypertrophy of the Kidneys, Rupture of the Kidney, Cystitis (Inflammation of the Bladder), Vesical Retention, Incontinence of Urine (Enuresis), Urinary Calculi (Gravel).

CHAPTER 22.
WOUNDS.
Unhealthy Wounds, Proud Flesh, Punctured Wounds, Injuries, Lacerated Wounds, Bruises, Ulcers and Fungous Growths, Bleeding Fungus.

CHAPTER 23.
DISEASES OF THE GENERATIVE ORGANS.
Difficult Parturition, Inversion of the Uterus, Mammitis (Inflammation of the Udder), Sore Teats, Orchitis Inflammation of the Testicles, Hydrocele (Water in the Scrotum), Sterility.

CHAPTER 24.
HOG CHOLERA AND SWINE PLAGUE.

INTRODUCTION.

HEALTH OF THE PIG.

Edmund Park says: "If we had a perfect knowledge of the laws of life and could apply this knowledge in a perfect system of hygienic rules, disease would be impossible. Hygiene is the art of preserving health. It aims at rendering growth more perfect, decay less rapid, life more vigorous, death more remote." So beautiful and comprehensive is this definition that it ought to be often repeated.

In dealing with this subject of health there are several things to be taken into consideration; this I will do as briefly as possible. First, we should follow nature's steps as closely as practicable, and should consider the condition of the pig in its natural haunts, and deprive it of as few of them as possible. The pig is an omnivorous animal and eats all. It is destined by nature to uproot plants and grope for food among the dropped acorns and other fruits of the forest, and Youatt says: "In point of fact the snout of the pig is its spade with which it roots in the ground for roots and earth worms." By putting an iron ring through the cartilage of its nose we thus deprive it of the power of searching for and analyzing its food, and by doing so we prevent it from getting substances which would be very beneficial for the maintenance of its health. To be profitable it is necessary to feed pigs more food than they could obtain in a natural state, in order to bring them to maturity as fast as possible, and this is done at the expense of the animal's health. Seeing that this has to be done, we ought to consider what kind of food is best to obtain this result and at the same time keep the animal in a vigorous condition. Yeo says that if an animal is in perfect health the pure alkaline blood circulating through the tissues of the body prevents the germs of disease from finding a suitable place to develop. Let us look for a short time at the physiological actions of some of the most important organs of the animal body, as we will then be better able to understand some of the causes of ill health. The stomach of the pig in its natural state is small and the intestines have great assimilating power. In this capacity the pig is ahead of all other animals, which accounts for its taking on fat so rapidly. By giving large quantities of food the stomach becomes distended, and in some cases, weakened so that it cannot digest the food properly and it passes out of the stomach in this condition into the intestines, where it acts as a foreign body, setting up disturbance, deranging the mucous membrane, leaving it in a condition favorable for the development of microbes and other germs of disease, the indigested portion will pass out as feces. The pig should be fed as much during the fattening period as it can digest and nothing more. This can be easily ascertained by examining the feces. The kidneys secrete the urine and other effete material, the result of the disintegration of the nitrogenous substances in the body; they require to be in a healthy, active state to perform this function, or blood poisoning is the result; if not blood poisoning, sufficient disturbance is caused to leave the animal liable to disease. The heart should be strong and vigorous in order to be able to propel the blood to all the tissues of the body to nourish them. The lungs should be strong, with large capacity to draw in oxygen and give off carbonic dioxide and other effete materials, in this way keeping the blood pure. The nerves which govern all parts of the body should be strong and active. This is largely accomplished by the kind of food we feed the animal.

What is the animal body composed of? The chemical constituents of the animal body may be thus classified: First, albuminous substances, characterized by the presence of nitrogen, carbon, hydrogen and oxygen. Second, carbo-hydrates and hydro-carbons, characterized by the absence of nitrogen and the presence of carbon, hydrogen and oxygen. Third, salts and water. In order to keep all the tissues of the body in healthy action and vigor, it is necessary to see that the animal gets a food which contains all these elements or to give a mixed diet which will combine to furnish the materials necessary. Food should be composed of nitrogenous portions called albuminates or flesh makers; hydro-carbons, or fat makers; carbo-hydrates, which are starch and sugar bodies, also fat producers. These are all necessary for the healthy development of the animal tissues. Let us see which of the various grains contain the substances mentioned:

	Corn.	Oats.	Peas.	Red Clover.
Water	13.9	13.5	13.8	16.7
Albumen	10.1	11.9	22.4	13.4
Fats	4.8	5.8	2.5	3.2
Carbo-hydrates non-nitrogenous extractive matters	66.8	57.5	52.3	29.9
Cellulose	2.8	8.1	9.2	35.8
Ash	1.7	2.6	2.5	6.2

These figures vary considerably, according to the ground on which the grains grow, whether it is rich or poor, cultivation, etc. The above table shows that oats and peas are more evenly balanced than corn. They are, therefore, the grains best suited for the growth and development of the tissues of the body, and also to keep them in a healthy state. When food substances are deficient in the albuminates and salts, the system is generally lowered in tone, and there is a tendency to the formation of "exudations," composed of imperfectly developed cells, which, in the great majority of cases, from the very beginning, are incapable of development into perfect entities, having only one potential quality, that of dying, and in so doing cause various derangements in the body, especially in the respiratory organs, producing tuberculosis and affections of the glands of the intestines. Oats also contain a nitrogenous alkaloid, called avenin, which possesses the property of acting as a nerve stimulant. It is on this account that horses largely fed on oats are so spirited. The salts or ash that these substances contain are all needed in the animal body in order that they will grow, and also support the system in older animals. Oats is the grain par excellence for the horse, and peas for the pig. Corn, alone, has not sufficient albuminates and salts and has too much starchy substance, which is converted into fat, and is therefore a grain which is not fit food for a young growing animal. It is necessary to feed other materials which contain albuminates to supply the deficiency of this material in the corn. And I am satisfied that the prevalence of cholera among pigs in the corn growing States is in a great part due to the feeding of too much corn. In Canada, where the pig is mostly fed on peas and oats and the refuse of wheat and rye, cholera is unknown. It is true there have been a few cases of cholera in Canada, but it has been mostly on the borders where it was supposed to have been brought over the river, and some years ago at Montreal, supposed to have been caused by feeding on distillery slops. Messrs. Lawes and Gilbert made a number of experiments on feeding in England and found that pigs fed exclusively on corn would frequently swell in the neck. They did not wish to discontinue the experiment, and therefore resolved to try the effect of putting some mineral substance in a trough within the reach of the pigs. They made a mixture of twenty pounds of sifted coal ashes, four pounds of common salt and one pound of superphosphate of lime. A trough containing this mineral mixture was put into the pen at the commencement of the second fortnight, and the pigs began to lick it with evident relish. From this time the swellings or tumors, as well as the difficulty in breathing, began to diminish rapidly, and at the end of a month had entirely disappeared. The three pigs consumed of the mineral mixture described above nine pounds during the first fortnight, six pounds during the second, and nine pounds during the third. This, although only a single experiment, shows, I think, that pigs may be fed on corn with impunity, providing that a compound of this or some other may be put within reach of the pigs. I would suggest the following: First, that we should avoid in-breeding as much as possible, as there is no doubt that it lessens the vitality of the offspring, leaving them in a condition liable to disease.

Second, that we select large sows, well developed and at least one year old. Third, that the boar should be of a smaller breed, compact, and of a vigorous constitution. This combination will insure strong, healthy offspring. Fourth, that the sow and boar should be fed on ground oats and bran mixed sufficient to keep them growing, but not too fat, as when they are too fat their vitality is lessened. They should have a small field to run in, separate, at some distance from each other. They should not have rings in their noses, but should be allowed to dig at pleasure, as they will find material in the ground useful for their health. If they should show signs of getting too fat, cut down their feed; on the other hand, if they are losing flesh, feed a little more. They should have a shelter from the sun in summer and a comfortable place to sleep in at night in the winter. They should have green clover in summer and dry clover hay in winter. Give them plenty of fresh water and a little salt mixed with their food. Pigs treated in this way will seldom have any ailment. Fifth, that having strong, healthy, young pigs to begin with it is necessary to feed them on materials that will keep up vigor and at the same time produce rapid growth. This can be accomplished by feeding them on ground oats or peas mixed with bran, and turning into a clover field if possible; if not, clover should be cut and brought to them. Milk of all kinds is useful. They should have a field to roam in, and after they are old enough the boars should be separated from the sows. The above food contains all the elements necessary for the growth and development of the pig. The bran, shell of the oats and the clover contain a large percentage of cellulose, and although the pig cannot digest more than half of this material, yet it is very useful, as it contains just what is needed to assist in forming the tissues of the body. Pigs fed as above will have all parts of their body well nourished and in a state of vigor to perform all the functions required of them to fortify the body against at least ordinary diseases.

Sixth, that too many pigs should not be kept together, as they are apt to sleep in the same place, and although it may be well ventilated, or even out in the open air, they are apt to breathe some of the foul air emanating from their bodies. No class of animals thrive well where numbers are kept together. When the time arrives to feed the hogs for market you will have a splendid foundation to begin feeding on; strong digestive and assimilating organs, which will be able to digest and assimilate large quantities of food. Corn can now be used with a little ground oats and bran with advantage and profit. I think that if this method were carried out, in a few years hog cholera would be a thing of the past.

There has not been the same attention paid to the treatment of the pig as there has been to that of the other domestic aimals. The difficulty in administering medicine to the pig and attending to the nursing of it is one of the drawbacks. Medicine has been usually given in the food, and when the animal is in condition to eat and can be separated from the others this method is very convenient; but putting medicine in a food where there are a number of hogs feeding together, cannot be done satisfactorily. In cases where it is necessary to give medicine by the mouth, the patient struggles so much that it often does more harm than good. When medicine has to be given to hogs, it is best to give it in the food, if they will eat it. If there are a number to be treated, each one should have its own allowance. When it is necessary to give it by the mouth, a piece of rubber hose should be put on to the neck of the bottle containing the medicine and tied firmly. The hog is to be then cast and secured either by being held or by tying. Then open the mouth with a piece of wood, introduce the rubber hose and pour the medicine slowly down. This is the safest and best way to administer medicine to swine.

The doses given in this book are for the adult pig. For pigs from six months to nine months, two-thirds; from three to six months, one-half; from six weeks to three months, one-third, and for pigs younger than this the dose should vary from one-sixth to one-twelfth.

DISEASES OF THE HOG.

CHAPTER I.

THE ORGANS OF MASTICATION.

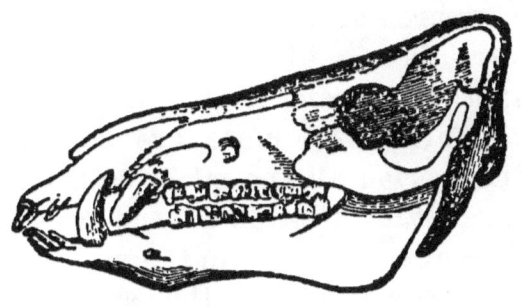

FIG. 138.

Skull of a Hog—showing the teeth.

Dentition of the Pig.—It is of great importance to have a correct knowledge of the appearance of the teeth so that we can determine the age of the animal. All those who are engaged in the management of show pigs, or are dealing in them, should study this subject. At birth, the young pig has eight teeth; four temporary incisors and four temporary tushes; about the tenth day appear the second and third temporary molars; at one month, four incisors are out, two in the upper and two in the lower jaw; about the sixth week, the temporary foremost molars are visible; at three months, two more are added to each jaw; at this period all the milk teeth are in position. Time is then allowed for the jaws and teeth to grow, and at six months, in the majority of pigs, a small tooth comes up on either side of the lower jaw, behind the temporary tushes, between them and the molars; and in the upper jaw, directly in front of the molars; at six months, the fourth molar appears through the gums; at nine months, the corner incisors are displaced and permanent ones make their appearance. The permanent tushes are also cut at this time, and the fifth molar on each side of both jaws makes its appearance. At one year the middle incisors are replaced by permanent ones, and by this time the tushes are of a considerable size; at this period the temporary incisors are shed and replaced by permanent ones; at eighteen months, in most pigs, dentition is complete, as the lateral incisors and the sixth molar are up.

Fig. 32.

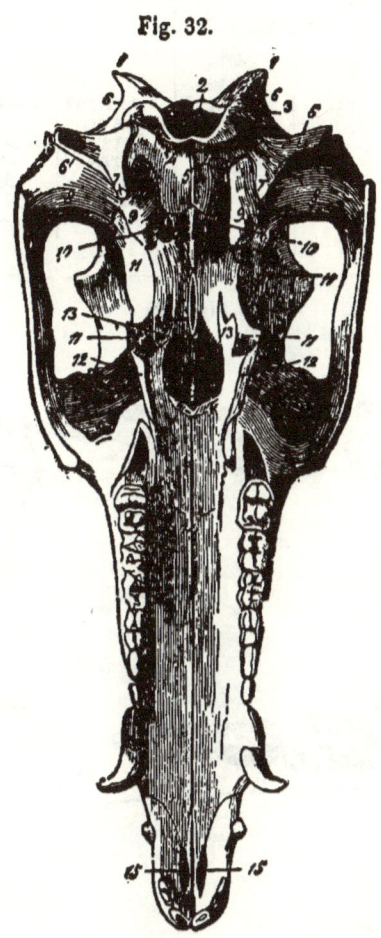

Prof. Simonds furnishes us with the following useful table:

24 DISEASES OF THE HOG.

	At Birth	1 Mo.	3 Mos.	9 Mos.	12 Mos.	18 Mos.
Foetal incisors	4	4	4			
Foetal tusks	4	4	4			
Temporary incisors		4 central	8 central and lateral	8 central and lateral	4 lateral	12 central, lateral, and corner
Permanent incisors				4 corner	8 central and corner	
Permanent tusks				4 cutting	4	4
Total in both jaws	8	12	16	16	16	16

Fig. 165.

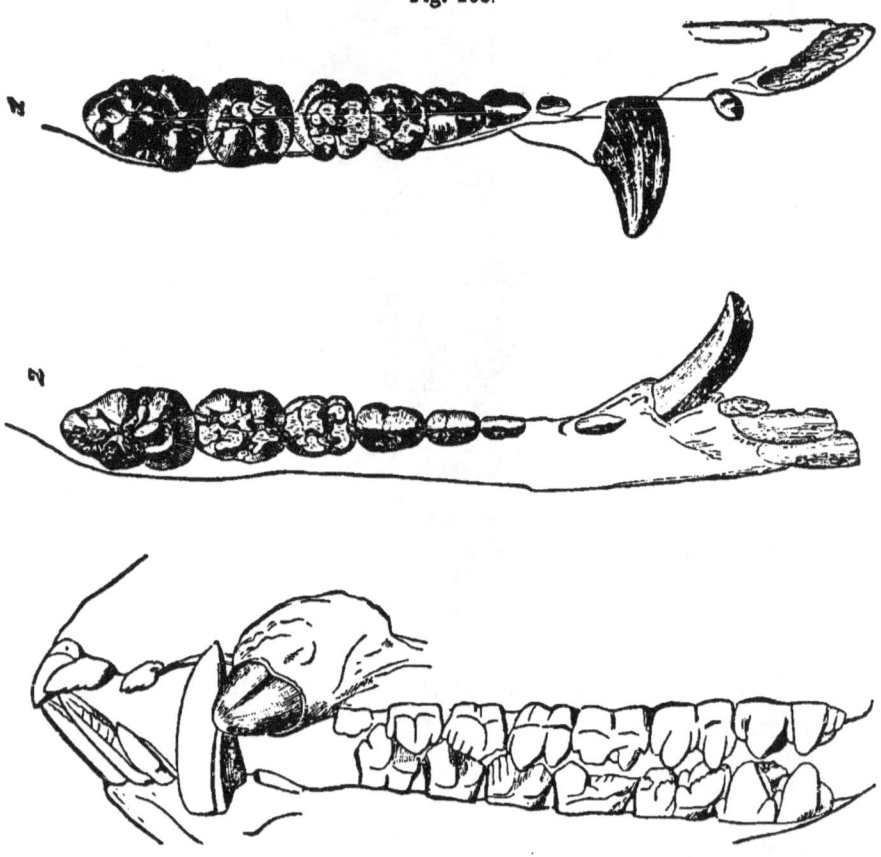

THE TEETH OF THE PIG.—

The mouth of the hog is large and the labial fissures extend far back; the upper lip is blended with the snout; the canine teeth are well developed, especially in the male; the lower tusks are long and curved outwards and upwards; the upper ones pass downwards and outwards, and they continue to grow during the life of the animal.

CHAPTER II.

DISEASES OF THE ORGANS OF MASTICATION.

Pigs, like other animals, suffer from teeth derangement, such as malformed teeth growing too long or turning inwards on the tongue. Mr. H. had a valuable sow which had difficulty in chewing and swallowing her food; she would chew a few times, then attempt to swallow, and the greater part of it would drop out of her mouth; there was also considerable dribbling of saliva; she lost flesh fast, and the owner called me to examine her. On opening the mouth, I discovered that the third molar tooth had grown long and either grown inward or accidentally gotten bent inwards and injured the tongue. I removed it with a pair of wolf's tooth forceps and applied a strong solution of alum water to the injured tongue and the animal improved rapidly. In another case a pig was suffering very much and had the symptoms of choking, or of having gotten something in its throat. The owner poured melted lard down, but it did no good. The animal would neither eat nor drink, but kept moving around, and at times would squeal. It kept on in this condition for three days, and as there was considerable saliva dribbling from its mouth, the owner began to fear it was mad. I was called to examine it and saw from the above symptoms that there must be some foreign substance either in its mouth or throat. I had the pig tied (it was nine months old) and the mouth opened, but could see nothing. I examined it several times with the same result, but being sure that there must be something there, I tried again and this time I saw a small white body sticking under the side of the tongue. I removed it with a pair of forceps and on examining it found it to be a temporary tooth. Both of these cases would have died in great suffering if they had not been relieved.

SPORADIC APHTHA.

This is vesicular eruption of the mouth. Although the pig is not so subject to diseases of the mouth as cattle, yet we occasionally see a case. The causes are usually local, such as the animal taking some irritating substance into its mouth. The symptoms consist of difficult prehension of food, the animal will take food into its mouth and let it drop out again; there will be a dribbling of saliva, and young pigs will squeal. On examining the mouth, clusters of white vesicles will be seen on the cheeks, lips and tongue. The treatment consists of dissolving half an ounce of borate of soda in a pint of water and applying a little of this to the mouth with a small mop. Another good remedy is equal parts of honey and vinegar, to which may be added a teaspoonful of carbolic acid to the pint, to be used as above. If ulcers appear they should be touched with a pencil of nitrate of silver, and if not better, touched again on the third day with the silver. Alum water is also useful, a tablespoonful of the alum to a half pint of water. Milk or thin oatmeal gruel should be given to the pig to drink.

There is a disease of the mouth known as "Gum Mouth," which is a form of Gloss Anthrax. This is a constitutional disease, locating itself in the mouth; it is accompanied by a low form of fever and is contagious.

Symptoms: A pig affected with this disease will refuse food, the eyes will be dull, ears lopped, saliva dribbling from the mouth, hot, dry nose, and the appearance of small pustules on the tongue and lips; the tongue will be coated with a fur and the edges will be dark-red and raw, and a very foul smell will issue from the mouth. These pustules break and discharge a very poisonous substance which very soon kills the pig. They should be cauterized with liquor hydrargyri nitratis, and a glass rod used to put it on. The strength should be kept up by giving four grains of quinine dissolved in two tablespoonfuls of whisky three times a day. Eggs and milk beaten up and a little whisky added to aid digestion should also be given in small quantities several times a day. The mouth should be washed out occasionally with a solution of slippery elm bark with a few drops of carbolic acid added to it. The pig should be kept in a clean, dry place, and given all the cold water it will take.

Glossitis (Inflammation of the Tongue) is a disease which may be seen in all our domestic animals, and is caused by injuries and by the animal getting too hot food or irritating medicines.

Symptoms: There will be swelling of the tongue, the animal will be salivated, and in bad cases the tongue will be protruded from the mouth; the swelling, if excessive, will cause difficult breathing, and if not relieved will cause suffocation; there will be a general derangement of the system; in some cases the epithelium (the scaly covering of the mucous membrane of the mouth) may peel off, leaving the parts raw and sore.

Treatment: In the early stages of the disease, examine the tongue if possible, sometimes this cannot be done on account of the swelling, if any foreign body is found remove it and bathe the mouth with cold water for half an hour several times a day and apply a little of the following lotion with a sponge: acetate of lead one dram and water one pint, shake up well and add one dram of tincture of opium. Give a purge, such as one to two ounces of sulphate of magnesium dissolved in half a pint of water, or if this cannot be taken, give a pill made of one-fourth of a grain of elaterium and one grain of extract of hyoscyamus. Feed on oatmeal gruel or eggs and milk.

GANGRENOUS INFLAMMATION. (GANGRENA ORIS.)

This is a disease which is usually seen in the young, especially those having a white skin. The probability is that it depends upon a peculiar morbid condition of the system and only wants some exciting cause to bring it out. The teeth seem to be in some way connected with it.

Symptoms: It is not usually noticed in the early stages. It begins most commonly by the appearance of white or ash-colored patches on the gums, most frequently below the lower incisor teeth, although it may occur on any part of the mouth; there is usually not much inflammation or swelling; in some cases where the inside of the cheek is the part affected, there may be so much swelling that it is observed from the outside, and the animal seems to suffer considerable pain, especially when it tries to eat; the patient often suffers from weakness. As the disease advances the slough spreads, and the external parts become hard to the touch; there will be a flow of saliva and the breath becomes fetid. I have seen a case of this kind where the complaint penetrated the deep structures and caused necrosis of the bone (death of the bone); the teeth became loose and the pig had to be destroyed. In the majority of cases, as the disease advances, the animal is unable to take food; an exhaustive diarrhea sets in and carries off the animal.

Treatment: In the early stages of the disease take sulphate of copper, half a dram to the ounce of water, and apply two or three times daily, so as to touch every portion of the diseased surface. Tincture chloride of iron, full strength, applied to it is also useful. Nitrate of silver, either in solution or pencil, is also good. When there is much fetor the mouth should be washed with a solution of carbolic acid, one to fifty. The little animal should be supported by brandy and eggs, one to two grains of quinine should also be given twice a day.

BLACK TEETH.

There has been a great deal said and written on this subject, but the great majority of the profession at the present day consider that black teeth is not a disease which is capable of causing death. I have studied black teeth and have never been able to connect it with any of the diseases affecting the pig, and in the majority of cases it seems to do no harm to the animal's health. I have no doubt, however, that pigs suffer as well as other animals from tooth ache, but I have never seen any cases where I could detect it. I am not able to account for the cause of black teeth unless it be natural for some hogs to have them. If it should be discovered that a hog has a decayed tooth and is apparently suffering from it, by all means have it removed, which can easily be done with a pair of wolf teeth forceps. I can advise swine breeders not to trouble themselves if they should notice some of their pigs having discolored teeth.

CHAPTER III.

STOMACH OF THE PIG.

Fig 139, Stomach of the Pig Inflated.—A. Cardiac portion. B. Its accessory cul-de-sac. C. Pyloric portion. D. Lesser curvature. E. Greater curvature. F. Oesophagus. G. Pyloric orifice.

The stomach of the pig is simple, although it takes on somewhat of a compound form, and to a certain extent performs the function of carnivora and ruminants. It consists of two portions: cardiac and pyloric; the latter is the smaller, but the divisions are marked externally by a much more distinct contraction. At the upper and left portion of the cardiac half is a small cul-de-sac. The oesophagus is infundibuliform in its termination. The mucous membrane, for the most part villous (velvety) in its structure, forms two folds, which extend from the cardiac towards the pyloric orifice, representing undeveloped oesophageal pillars and canal. The gastric juice of the hog contains the same ferments as are found in the secretions of other mammals. The secretions from different portions of the stomach differ; that obtained from the greater curvature contains more mucin, more acid and more ferment than that from the other portions, while the secretions from the oesophageal portion are free from ferment. It is found that the conversion of starch into sugar continues in the stomach of the pig, as the food remains alkaline in the cardiac end; but as the food moves on and comes in contact with the acid gastric juice it ceases. The saliva of the pig is very active, and its action continuing after it has reached the stomach gives the pig very great power of digesting starchy food, and on this account the pig takes on fat more rapidly than other animals, while being fed on cereals, especially corn. It has been noted by experiment that flesh takes a much longer time to digest in the stomach of the pig than it does in that of carnivora. The pig does not masticate vegetable matters as well as herbivorous animals, so that they are less constituted for the extraction of nutritive principles from it; therefore, although meat and vegetables are useful as articles of diet, they are not so profitable for feeding purposes as grain. It is claimed that the pig is capable of digesting fully fifty per cent of cellulose. Figure 140. The intestines in general resemble those of the ruminent. The caecum resembles that of the horse. The intestines are not nearly so sensitive and therefore are not nearly so liable to disease as those of the horse; they are short and the absorbent glands are numerous and active.

CHAPTER IV.

DISEASES OF THE STOMACH.

Indigestion.—The pig, like other animals, suffers at times from derangement of the stomach. If fed for a long time on one kind of food it is likely to be affected with indigestion, loss of appetite, dullness and loss of flesh, and this condition is favorable for the development of worms or ulceration of the stomach. It is therefore necessary, in order to keep a pig in good health, to give it a mixed diet or complete change of food for a few days. There are a number of diseases in other parts of the body which are caused by a faulty digestion, such as diarrhea, vomiting, lung and skin diseases and a number of others.

Symptoms of Indigestion: The appetite is usually more or less impaired, and sometimes wanting altogether, constituting "anorexia;" in other cases again there is a morbid craving for stuff that they would not touch in health; they will come up to the trough, take a few mouthfuls, then leave off; in some cases the pig will press its nose against the ground and may whine or squeal; sometimes it will vomit up a thin, sour-smelling liquid mixed with a little half masticated food; the bowels may be constipated or there may be diarrhea; in chronic cases there is often a cough and the pig may suffer from headache, or it may stagger from giddiness and even fall over; in young pigs it causes fits. The animal in this condition will not thrive or grow, but usually loses flesh and sometimes becomes emaciated with wasted muscles and a sunken abdomen. The pulse, in some cases, is quite natural; in others it is somewhat increased in frequency or is irregular; there may also be fever and scanty, high-colored urine.

The causes of indigestion in the pig are want of exercise and too much food, or food of a poor quality; hence it results in weakening the stomach. To prevent this the pig should be allowed to run at large in a field; especially is this the case in the young pig, as it requires more exercise than the adult. It should be regularly fed on nutritious food and not too much of it.

Treatment: If the pig is constipated give from one to two ounces of epsom salts and a teaspoonful of ginger, dissolved in half a pint of water, at one dose. If there is diarrhea give from one to two tablespoonfuls of castor oil or from a dessert to a table spoonful of tincture of rhubarb. After the physic has operated give a teaspoonful each of tincture of gentian and ginger at a dose three times a day; or if the animal will take a little food give from five to ten grains of sulphate of iron and a teaspoonful of ground anise at a dose in the food twice a day. If the animal is troubled with vomiting give one to two drops of the wine of ipecac; or five drops of carbolic acid in a little sweetened water will be found useful; ten drops of nitro-muriatic acid in a little water given twice a day is also good.

GASTRITIS (INFLAMMATION OF THE STOMACH.)

This disease is usually seated in the mucous membrane and the sub-mucous areolar or connective tissue; but in some severe cases the muscular tissue is also involved.

Acute gastritis is not a common disease in the hog, and it is not often seen as an independent affection, but is more frequently associated with some other disease. Wood says: "Few organs resist so firmly the ordinary direct causes of inflammation as the stomach, and few are so readily affected through the sympathies."

Causes: Inflammation of the stomach is usually caused by the pig eating some indigestible food which sets up irritation, or by caustic medicines, or in some few cases by rheumatism.

Symptoms: There will be vomiting, great pain, restlessness, the pig moving about almost constantly from place to place, and occasionally squealing; it will refuse food, but may be thirsty; the substance thrown up will be, first, the contents of the stomach, afterwards, bile or mucus often tinged with blood; the end of the nose is dry, and if the tongue is examined it will be found to be coated with a whitish fur; the bowels are usually constipated; the animal breathes fast, and the pulse is full and frequent at first; the skin is dry, the urine high colored, and there is sometimes a hard, dry cough. As the disease advances these symptoms become intensified; debility and restlessness increase; delirium sometimes takes place, or the animal may become partially paralyzed and soon sink and die. On the other hand, if the disease should take a favorable turn, the vomiting will become less frequent or cease altogether; the animal becomes quiet, lies down and may go to sleep, and after a while may be looking for food. In some cases of gastritis caused by irritant poisons, the animal may die in a very short time; but in the majority of cases the pig will live for twenty-four hours to four or five days.

Treatment: If possible, find the cause. If it should be a strong acid give carbonate of soda or lime water; or if nothing else is at hand scrape some of the plaster or whitewash off the wall, mix it with water and give it as soon as possible. On the other hand, if it should be by an alkali, give vinegar, then give flaxseed water, barley water, or gum arabic dissolved in water; of if nothing else is on hand give milk. The animal should always get from fifteen to twenty-five drops of tincture of opium in a little water every hour, or one to two grains of powdered opium and a half grain of calomel. If from indigestible food, give one to two ounces of castor oil; follow this with linseed tea or gum arabic dissolved in water, and the opium as before. When the vomiting has lasted for some time it ought to be checked if possible, and this is best done by giving from ten to twenty drops of the medicinal solution of prussic acid in a little water, or one to two drops of the wine of ipecac. Allow the pig all the cold water it will take. After the acute stage is over, the pig should get a little new milk, with a little whisky or brandy in it, several times a day. Care must be taken not to allow the animal to have much food for a week or ten days; a little oatmeal made up with boiling water and mixed with milk will be the best food.

Post Mortem Appearance.—It was at one time thought that redness indicated that the stomach was inflamed, but mere redness may be present after death and the stomach have been healthy. In cases of true inflammation the mucous membrane will be very much swollen and congested. In some cases I have found swelling so great that it had completely closed the cardiac opening into the oesophagus. In such cases the animals cannot swallow, there is considerable infiltration of serum mixed in the tissues, which are usually easily broken down and will be either almost black or of a yellow tinge.

MEAT BRINE.

Salt in moderate quantities promotes digestion and the general health of the animal; but when taken in too large quantities it deranges the stomach and bowels, causing the formation of gases, diarrhea, vertigo, convulsions and paralysis, and death in eight to twenty-four hours. It also causes acute inflammation of the stomach or bowels, or both. I have treated pigs which got too much salt brine, causing the above. The mucous membrane of the stomach, and in some cases part of the intestines, were found, after death, highly injected and swollen in patches and of a dark-red or a greenish-yellow color. Pigs should not be allowed to get brine of meat, unless in very small quantities, and then it should be boiled. In cases where saltpetre has been used in conjunction with the salt in curing meat the brine should not be used. The symptoms are those of gastritis. Treatment will depend on the nature of the case. If seen early and the pig has not vomited, mix a tablespoonful of mustard in half a pint of hot water and pour it down. If the animal does not vomit in fifteen minutes repeat the dose. Then give one to two grains of powdered opium in a little sweet oil every two hours to relieve the pain. The animal will be very thirsty and should get water in which barley or slippery elm has been put. If there is severe diarrhea add five grains of acetate of lead to the opium; if there are convulsions give bromide of potassium in two to four dram doses, dissolved in water, every two hours. In cases of paralysis give a tablespoonful of spirits of nitrous ether mixed with half the quantity of aromatic spirits of ammonia in a little water every two hours. If there should be constipation give from one to two ounces of castor oil.

SIMPLE ULCERS OF THE STOMACH.

I have met wth several cases of ulceratoin of the stomach of the pig as an independent disease and in conjunction with other diseases, such as hog cholera. In one well marked case, a fine sow, which had been in a thrifty condition until she was one year old, was noticed to vomit occasionally, and seemed to be somewhat uneasy after eating; this continued for several months and she began to lose flesh; and being a valuable sow I was called to see her. By this time some blood was mixed with the food in the vomit, and I diagnosed the disease to be either ulceration or cancer. I examined some of the material vomited, but could not get any satisfactory results from it.

Symptoms: The appetite is variable, in some cases it may not be much affected. The animal will begin to eat its food with an apparent relish and all at once it will stop feeding, leave the trough, apparently in pain. It may vomit or seem to be trying to do so, something between a cough and an effort to vomit, and there may be only eructions of gas. The animal soon gets into an unthrifty condition; the bowels are usually confined and the urine scanty and high colored; the pulse and breathing are not affected in the early stages of the disease. If the animal is not relieved it gradually becomes worse, vomits up nearly all its food, seems to be in much pain, and the contents of the stomach are usually mixed with blood. There is no other disease that can be mistaken for this one except inflammation, which is very short in its duration, while ulceration may last for nearly a year before it destroys life.

Treatment: There are a number of remedies which can be used in the treatment of this disease. Subnitrate of bismuth in ten grain doses three times a day is very useful; this should be given on an empty stomach. Half a grain of nitrate of silver and half a grain of opium is a very valuable remedy; it is best given in pill and should be administered three times a day before feeding. Small doses of sulphate of iron are useful. One grain of mercury in the form of blue mass and one-sixth of a grain of ipecacuanha made into a pill and one given three times a day for a week is good.

Post Mortem Appearance.—The ulcers are usually found in patches of various extent and the mucous membrane surrounding them is swollen and of a dark-red or bluish color. The ulcers are of various shapes; some of them are pitted, others are filled up with a grayish-brown substance. In the majority of cases they are hard to the touch or to cut with the knife; sometimes they are soft. A number of these ulcers seem to join together, forming a patch one to two inches long and half an inch wide. The root or base of the ulcer usually extends through the stomach, forming a hard, bluish tumor; and it is said that sometimes they slough out, leaving an opening through the walls of the stomach, although I have not seen such a case.

VOMITING.

Vomiting is the act of ejecting material from the stomach. It is accomplished in two modes. Simple vomiting is merely regurgitation or eructations effected by contractions of the stomach, sometimes assisted by the voluntary contractions of the diaphragm. Second form, which is the result of the combined actions of the stomach, diaphragm and abdominal muscles.

Causes: The most common cause of vomiting is inflammation or irritation of the stomach. Substances in the form of food which are not readily digested become sour and irritate the stomach. Diseases of the abdnominal viscera are very apt to cause it also. There are conditions of the nervous system, produced by various causes, which dispose to that cerebral action essential to vomiting. Injuries to any part of the body which produce shock or exhaustion will cause it. Disorders of the brain are often accompanied by violent vomiting. Sometimes pigs may appear well and eat heartily and in a few minutes afterwards vomit; this form is usually the result of ulcers and can usually be prevented by putting a few drops of carbolic acid in the food, five drops are usually sufficient.

Treatment: It is obvious that we should find the cause, as upon the condition depends the value of the remedy. If from eating indigestible food, give a teaspoonful of aromatic spirits of ammonia. If from inflammation or irritation, give from one to two grains of opium in pill, or from one to two drops of the wine of ipecacuanha. If it is rejected by vomiting before it has had time to act, mix from twenty to sixty drops of tincture of opium in a little thin starch gruel and give as an injection per rectum. If this does not stop it give hypodermically one-sixth of a grain of morphea in twenty drops of pure water. One drop of creosote at a dose and repeated if needed is sometimes very useful. Some place great confidence in chloroform in doses of from five to twenty drops. This may be tried if the others fail.

MORBID APPETITE.

There are two forms of this derangement: first, an animal may eat enormous quantities of food and still not be fat; second, it may eat unusual substances. In the first case the animal should be allowed only a reasonable quantity of good food for several weeks until the stomach becomes accustomed to it. The second form is usually called depraved appetite (Pica). In this disorder there is a desire, which seems to be irresistible, for substances wholly unfit for food. This is often a habit, but may be caused by some deranged state of the stomach. The desire for earths, lime, stone, etc., would indicate that the animal's stomach was in an acid condition.

Treatment: In the first case regulate the food so that the animal cannot get too much. The stomach and bowels are usually in a weakened condition and require toning up, which is best done by giving from five to ten grains of sulphate of iron and a dessert spoonful of aniseseed at a dose in its food twice a day for a month, and by this time the animal will usually be cured. If not, it would be wise to kill it, as those morbid feeders never do well. In the second, put the animal in a place where it cannot get at the material it eats, and give it a dose of epsom salts, one to two ounces, then give the above tonic. If the animal has been in the habit of eating earth, give it bicarbonate of soda in its food, one teaspoonful at a dose, combined with the tonic.

HEMATEMESIS.

Bleeding from the stomach is usually attended by vomiting of blood, but not always, as the blood may have been swallowed, then vomited, without any hemorrhage direct from the stomach. It is not common in the pig, although I have seen a few cases.

Causes: Injuries in the region of the stomach, such as a kick from a horse. I have had a case from this cause. Hard substances that the animal had swallowed, caustic substances introduced into the stomach, violent straining in vomiting and from inflammation and ulceration. It is also produced from diseases of the liver, spleen and other organs.

Symptoms: Hematemesis may be preceded by loss of appetite or it may come on suddenly, which is the case when it is caused by violence. Hemorrhage may take place in the stomach and pass off by way of the bowels, which often happens when the quantity is too small to induce vomiting, the feces in this case are usually black. When we suspect hemorrhage the mouth and nostrils should be examined to find their condition. There is no difficulty in discriminating between hematemesis and hemoptysis (bleeding from the lungs). In the former, the blood is usually dark in color and coagulated and mixed with the contents of the stomach and is discharged by vomiting. In the latter, it is bright red and frothy, never coagulated, frequently mixed with mucus, and brought up by coughing. Death is not the usual result, although the disease may be speedily fatal. In one case where a pig was kicked by a horse, causing hematemesis, the pig died in a short time. I made a post mortem and found the stomach distended with blood. In this case the animal did not vomit. So it may be that many cases may take place and kill the animal and the cause of death not be known.

Treatment: Acetate of lead from one to two grains and opium one to two grains, given at a dose and repeated every two hours, is a most efficient internal remedy. If this should be rejected give lime water and milk to quiet the stomach, then give fifteen drops oil of turpentine and from fifteen to twenty-five drops tincture of opium, repeat every two hours; if this should be rejected give a teaspoonful of tincture of opium in a little gruel as an injection. The fluid extract of ergot

46 DISEASES OF THE HOG.

of rye in teaspoonful doses may be tried. If the animal is very weak give from one to two teaspoonfuls of aromatic spirits of ammonia in a little cold water, repeat every hour if needed. After the hemorrhage has ceased feed on milk and eggs with two tablespoonfuls of limewater or two or three drops of carbolic acid in it. Feed very sparingly for a few days.

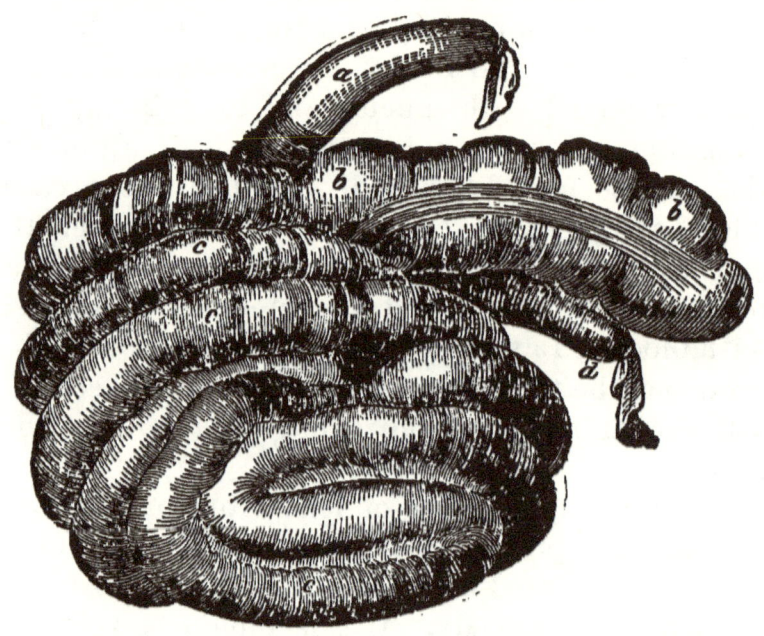

FIG. 140.

Cecum and colon of a hog—inflated. a, Ileum; bb, Cecum; ccc, Colon; d, Rectum.

CHAPTER V.

DISEASES OF THE INTESTINES.

Enteritis (Inflammation of the Bowels.) Enteritis, strictly speaking, means inflammation of any part of the bowels, but it is seldom that the bowel is inflamed throughout its entire length. This is a very fatal disease in all animals, especially in the pig. The causes are colds, injuries, irritating substances in the food, obstinate constipation or diarrhea. It is also caused by drinking bad water and sometimes is a result of other diseases.

Symptoms: A pig affected by this disease refuses food, is thirsty, is very restless, has a dull appearance and suffers much pain will lie down, roll on its side, get up, move around and grunt or squeal the mouth is dry and of a purplish color, and if the pig has a white skin it will be red in patches, especially on the belly; there is a great tenderness of the walls of the abdomen and the animal will squeal or moan if pressed upon; diarrhea is usually present and the discharge from the bowels is apt to be followed by an increase in the pain; but after a few minutes seems to be mitigated somewhat. The discharge may be very frequent and tinged with blood, occasionally it is of a dark or green color, being charged with bile; sometimes there is flatulent distention of the bowels, the temperature is usually from 104 to 105 and the pulse at first is full and soft and very frequent, from 120 to 150 per minute; as the disease advances the pulse becomes weak and almost imperceptible; the breathing is short and fast and the animal may have shivering fits; there is often severe vomiting, which is very distressing to the patient. This disease in the pig usually lasts from three days to one week and is caused most frequently from mortification and collapse.

Post mortem appearance: The mucous membrane is thickened and gangrenous and often there will be ulceration, which might lead one to call it "Hog Cholera." The ulcers often penetrate the walls of the bowel, and if a number of them should coalesce under such circumstances a slight force of pressure is sufficient to cause rupture, allowing the contents of the bowel to pass into the abdominal cavity. I have frequently seen cases of this sort. Sometimes there will be a sloughing of the mucous membrane, caused by effusion between it and the muscular coat. Cases have occurred in which several feet of the mucous membrane had become detached. Portions of false membrane are occasionally observed adhering to the surface of the mucous membrane.

Treatment: When there is diarrhea present give from one to two ounces of castor oil with fifteen to twenty-five drops of tincture of opium in it this will clear out the irritating secretions or accumulations and the laudanum will assist in relieving the pain. If there is constipation epsom salts and manna will be found useful. After the physic operates give one-fourth of a grain of calomel and one grain of opium made into a pill three times a day. If the fever is high with a strong, fast pulse, give one to two or three drops of the fluid extract of veratrum viride in a little water until the pulse is reduced in force and frequency. The pig should get quantities of linseed tea or gum arabic, which will soothe the irritated membrane. If the diarrhea should continue, mercury with chalk should be substituted for the calomel. If the pain continues very severe, the dose of opium should be given larger or oftener. The animal should be kept in a dry, comfortable place with plenty of straw to lie down on, and solutions of arrow-root or sago with milk should be given as food to keep up the strength. External treatment is not practicable. As soon as it becomes convalescent feed on oatmeal and milk in small quantities for a week or two.

DYSENTERY.

Dysentery is an inflammation of the mucous membrane of the large intestine, especially the rectum, characterized by severe tenesmus (straining) and the passage of small quantities of mucous or bloody feces, accompanied by pain.

Causes: Eating decayed vegetable matter or vegetables not easily digested, putrid animal substances and constipation; lying in cold, damp places at night is a very common cause; it is also caused by a congested state of the portal circulation and a sluggish condition of the liver, often seen in fat pigs, drastic purges, worms, and a sequel of adynamic diseases. A number of pigs on the same farm may be attacked at the same time without any apparent cause. In such cases it must be of a miasmatic nature and often takes on a typhoid form.

Symptoms of acute dysentery: This is not a common disease of the pig, although I have seen several outbreaks supposed to have been caused by bad water. It is usually preceded or followed by general uneasiness, dullness, impaired appetite, with stiffness in moving; there is more or less pain, as the animal whines; there may be constipation or diarrhea, the passages are usually lumpy at first, and very frequent; after the first few evacuations what is passed is of a whitish mucus or mucus mixed with blood. As the disease advances there will be shreds of the mucous membrane passed or small masses of coagulated matter. There is usually a good deal of flatus passed and relief follows for a time. At first the discharges have little smell, but after a time they become very offensive. There is always fever, except in very mild cases. Generally the disease takes a turn between the sixth and tenth day and the patient recovers; or sometimes it becomes worse at this time, which is indicated by great depression, the straining ceases to a certain extent or altogether, the animal lies down and can hardly be induced to rise, the skin is cold, the breathing is fast and short and the pulse imperceptible, the body may be covered with purple spots and the animal will die in a few hours. This disease is often taken for "hog cholera."

Post mortem appearance: In cases of death from dysentery there are always signs of inflammation of the mucous membrane of the rectum and lower portion of the colon, although the inflammation may extend much farther up, as I have found it in the small intestine. The membrane is much thickened and of a dark red color, and there are usually ulcers. These ulcers may be single or confluent. The single ones have abrupt edges and are often covered with a concrete exudation and somewhat resemble a slough. In some cases coagulated lymph will be found coating almost all the membrane; in some cases there will be real gangrenous sloughs. Very often in protracted cases the liver will either be in a congested state or of a dirty gray color, very easily torn, and the gall bladder distended with gall. The spleen may be enlarged and congested. In the majority of cases the lungs appear healthy. The kidneys are often congested and there is considerable effusion in the abdominal cavity with some peritonitis. The lymphatics are usually much enlarged and soft.

Treatment: The most efficient remedy in dysentery is a good physic, as it cleans the bowels and thereby we get rid of irritating secretions and also diminish congestion of the portal circulation; but it is necessary to be careful in selecting cathartics, as drastic ones may do more harm than good. If there has been diarrhea for some time it is likely that it has at least removed all feces, and a strong purgative will not be necessary, still the liver and portal circulation require stimulating. It is generally best to give from ten to fifteen grains of calomel and follow this in six hours with a dose of castor oil, say from one to two ounces, or if there should be much fever give instead of the oil one to two ounces of epsom salts or the tartrate of potassium and soda. The compound powder of jalap I have found very useful in from thirty to forty grains at a dose. After the physic operates give one grain of opium, two grains ipecacuanha and two grains of calomel, made into a pill, every two hours until the pig is easier or sleepy. Acetate of lead ten grains, opium one to two grains, given three times a day is very useful in some cases. In the advanced stages of the disease nitro-muriatic acid from five to ten drops, laudanum fifteen to twenty-five drops, given in a little mucilage three times a day or oftener if necessary is good. Oil of turpentine given in emulsion in from ten to twenty drops, and laudanum fifteen to twenty drops, given three or four times a day is also good. If the animal is very weak give two tablespoonfuls of good whisky with two to four grains of quinine in it three or four times a day. There are a number of other remedies which can be tried, but those given above are among the best. Injections are sometimes serviceable. Tincture of opium one teaspooonful, acetate of lead twenty grains, mixed in a little starch gruel and repeated every three hours is of service. Give the pig starchy food, milk and eggs. There is occasionally a chronic form of this disease, but the treatment will be the same as second stage of the acute form.

DIARRHEA.

When an animal is affected with a discharge of liquid feces it is called diarrhea or scours. This affection is rather a consequence of certain pathological conditions than itself a disease. The conditions which cause this derangement are various and at times even opposite; a simple increase of the peristaltic action may produce it without any other cause. It is often the result of a great excitability of the intestines, causing a much stronger impression than they are accustomed to in health, or from an increase in the amount, especially if it is of a stimulating character, or the introduction of irritant food or food that undergoes fermentation rapidly. It is often the result of some effete material in the blood or from increased secretion from the liver or pancreas; these substances often cause irritation sufficient to cause diarrhea. Debility of the mucous membrane may allow the elimination of fluids into the bowel, and have the effect of causing more contraction of the muscular coat. Young pigs are very frequently attacked by diarrhea from some substance the mother has eaten. It is also the result of dentition. Poverty of the milk given by the sow, improper food, irregular feeding, cold and damp sties, sudden changes of temperature, green food given to the sow with litter when they have been too long deprived of it, not infrequently cause it.

Symptoms: In simple diarrhea there is a discharge of liquid feces without any constitutional disturbance. The feces may be passed without any apparent pain or inconvenience to the animal. In other cases the discharges are very frequent and painful, which cause the animal much distress; this form is usually accompanied with fever, quick pulse, fast breathing and loss of appetite and a tucked-up appearance of the abdomen, and the animal very soon becomes exhausted. There is a form of diarrhea which sometimes will be seen in the pig, caused by derangement of the liver, and may be of two kinds: first the form which is caused by an increase in the secretions of bile; the feces are liquid of a bright yellow color, at other times they will be green from the action of acid in the bowel, and the passage is accompanied by severe straining and pain and usually vomiting. Second form differs from the first in that the passages are black or brown and of a very unhealthy appearance and foul-smelling. In these cases the appetite is usually impaired and there is considerable fever. This form of diarrhea is often mistaken for "hog cholera." All forms of diarrhea, if not attended to, will soon so reduce the pig that it may die.

Treatment: The treatment of this disease must vary according to the nature of the derangement. When the complaint simply depends upon increased peristaltic action it yields easily to a dose composed of fifteen to twenty drops of tincture of opium and the same quantity of spirits of camphor in a little water, repeated in two hours if not relieved. The cause should also be removed if possible. If it is caused by change of food or too much food give from one to two ounces of castor oil and from one to five drops fluid extract of belladonna; after the physic operates if the diarrhea is not checked give a few doses of opium and camphor usually one dose of purgative medicine is enough, as harm is often done by purging too much. If the pig has been purged for some time before being treated it will not do to wait the action of the physic, but give a dose of astringent medicine as soon as possible, such as tincture of catechu one dram, tincture of opium twenty to twenty-five drops, give this in a little starch gruel and repeat in two hours if not checked. If it is caused by bile derangement give from two to four grains of calomel and twenty grains of rhubarb in a little gruel; follow this with small doses, such as one-sixth of a grain each of calomel, opium and ipecacuanha in a pill every two hours. After the physic has operated if the diarrhea does not stop give opium one grain, tannic acid ten grains, in gruel repeat in two hours if needed. When the diarrhea is accompanied by very large watery passages and the pig is weak give opium one grain, acetate of lead five grains repeat this in two hours if needed. In the chronic form of diarrhea tonics and stimulants are called for, such as turpentine in from fifteen to twenty drop doses three times a day, or sulphuric acid in from five to ten drops three times a day. Besides this give a teaspoonful each of tincture of gentian and ginger in a little water three times daily. Sulphate of iron in from five to ten grains in a little food is also useful. Milk diet with two tablespoonfuls of lime water in it is very good. Well boiled flour gruel mixed with milk is very fine. For older pigs dry, whole grain of any sort will answer.

COLIC.

This is not a common disease in the pig, but we find a case occasionally.

Causes: It is caused by changes of food and food of an indigestible and fermentative nature, cold, lying in wet beds, from constipation and obstructions.

Symptoms: The pain of colic occurs usually in paroxysms with an intermission of ease. The pig is restless, shifts from place to place, sitting on its haunches, twisting its head from side to side, getting up, lying down, grunting or squealing; if the pain is severe the muscles of the abdomen are hard and may be rigid; pressure on the abdomen will ease the pain and the pig will sometimes lie flat on its belly for a few minutes. The pain often abates suddenly or it may last for hours, although it usually yields to proper treatment. If possible find the cause. If it is from constipation give a dose of castor oil and from fifteen to twenty-five drops of tincture of opium in it. If the pain is very severe give twenty-five drops of the essence of peppermint or spearmint dropped on sugar, or a teacupful of effusion of ginger or from one to three teaspoonfuls of the compound tincture of lavender. One to two teaspoonfuls of the camphorated tincture of opium will usually be still more efficient; any of the above should be repeated at intervals of from one to two hours. If it should be very persistent give one-fourth of a grain of morphine hypodermically. After the acute pain has subsided give the pig a dose of epsom salts to clean out the bowels.

CONSTIPATION.

This term is applied to a condition of the bowels in which the passages of feces are less frequent and of smaller quantities than normal. If this condition of things lasts for a time it will impair the health.

Causes: Sows and young pigs that have been too highly fed are often troubled by constipation. Pigs which are confined in houses are often the victims of constipation, as exercise is needful for the proper performance of the functions of the intestines. Food that does not stimulate the bowels to contraction, torpidity of the nerves of the intestines, and scant secretions. Obstructions to the bowels and mechanical impediments will be described under the head of obstructions.

Symptoms: The animal is uneasy, moves about, strains and may pass a small quantity of hard feces, there will be distention of the anus, the lump will frequently be covered with white or bloody mucus. If the accumulation should be up the bowel the animal may pass only mucus or bloody mucus, which might be mistaken for dysentery. In a short time if not relieved the animal becomes dull, the appetite will be impaired, and in some cases vomiting will occur, with severe sickness, which may end in death in a few days.

Effects: Constipation, besides causing irritation, inflammation, distention, ulceration, gangrene and piles, deranges the neighboring organs by the pressure of the accumulated feces. It impedes the circulation, causing congestion of the various organs and affecting the heart, brain, liver and skin. A great number of skin diseases are the result of constipation.

Treatment: Give from one to two ounces of castor oil this is a useful purge, being mild, sure and quick in its action. Sulphate of magnesia and other salines, on account of their causing a great increase in the secretions, are very efficient when there is an accumulation of hardened feces. Senna tea combined with salts increases their activity.

Cases may occur in which a drastic purge will be necessary; in such cases give from two to three drops of croton oil in a little castor oil or sweet oil. There are a number of other purgatives which are useful. The compound cathartic pill is very good, dose one to two pills. Injections of soap and warm water should not be neglected. After an attack of constipation the bowel is more or less weakened and the animal should have a stimulating tonic, such as five to ten grains of sulphate of iron and two to three grains of nux vomica at a dose three times a day in its food for a week. Young pigs fed on skimmed milk should have a little boiled flaxseed mixed in it, it is very nutritious and will prevent constipation. It is also good for grown pigs. A teaspoonful of white mustard seed and a little hardwood charcoal is a preventive.

OBSTRUCTIONS TO THE BOWELS.

This term is applied when some mechanical impediment obstructs the passage of the feces.

Causes: An accumulation of hard, impacted feces, the accumulation of hard substances such as coal cinders. I have met with several cases of this kind in which a farmer had lost a pig and had an idea that it might be "hog cholera." On post mortem I found a portion of the ileum near its entrance into the cecum filled with cinders and that portion of the bowel swollen and black, and I have no doubt but that hundreds of pigs die from this cause. Solid concretions sometimes form in the bowel. There are also strictures which take place, tumors, and in some cases organized bands across the bowel, originating likely from the process of inflammatory adhesions of the surface of the mucous membrane in consequence of the exudation of coagulated lymph and a subsequent separation of these surfaces, before the lymph has become quite consolidated, so that it is drawn out in apparently interlacing cords. Twisting of the bowel is still another cause of obstruction, invagination of the bowel is a frequent cause, strangulation by the passing of a portion of the bowel through a rent in the diaphragm, mesentery, etc.

Symptoms of obstructions: It may come on suddenly, and in such cases there is great restlessness, vomiting, straining to pass feces, with the effect that the abdomen will swell and the animal show all signs of inflammation and soon die. In other cases it comes on slowly, the animal strains but cannot pass feces; this continues in spite of purgatives and gradually goes on until the pig dies in great agony.

Treatment: Find the cause if possible but this is usually not easily done in the pig. First give a dose of castor oil and if this takes no effect give salts and senna or two ounces of castor oil and three drops of croton oil. When the pain is severe give one grain of opium at a dose every two hours or one grain of opium and half a grain of calomel every two hours. Injections of soap and warm water should be given every hour. If the vomiting is severe give a drop or two of wine of ipecac. Physostigmine in doses of one-tenth of a grain and a sixth of a grain of pilocarpine dissolved in a little warm water and administered hypodermically will sometimes overcome obstructions. Coal cinders should never be given to pigs. Charcoal and wood ashes are safe, or better still, twenty pounds of sifted coal ashes, six pounds common salt and one pound superphosphate of lime, mix these well together and put into a trough in a convenient place so that the pigs can get at it when they want it.

STRICTURE OF THE RECTUM.

When there is great straining and difficulty in evacuation stricture of the bowel may be suspected, and it is only after an examination that the cause is found. There are several things which may happen to the bowel which would prevent wholly or partially the evacuation of the feces. First, thickening or other organic derangements of the coats of the bowel; second, prolapsus of the rectum; third, hard tumors and spasmodic stricture.

Symptoms: This affection is usually considerably advanced before it is noticed, when there will usually be constipation, with severe and painful efforts to evacuate, and nothing but a small quantity of mucus will pass. When these symptoms are present no time should be lost in ascertaining the nature of the derangement. The finger should be oiled and introduced; if the cause is beyond the reach of the fingers a bougie should be used. When it is discovered what the ailment is it should be removed if possible; if this cannot be done the animal should be destroyed. If it is from thickening of the coats, and not too far up, the bowel should be dilated and the enlarged part rubbed with a strong tincture of iodine twice a week. If it should be a tumor either hard (scirrhus), soft, or polypus, it will have to be removed by the knife or ligature. It is a dangerous operation, but it is a case of life or death and ought to be tried. Spasmodic stricture is best overcome by the local use of the fluid extract of belladonna and tincture of opium applied to the inside of the rectum once or twice a day. The pig should have a dose of physic, epsom salts is the best, and the feces kept soft by feeding on laxative food.

HEMORRHOIDS OR PILES.

This is a term applied to soft tumors which are easily made bleed; these tumors are found in or about the anus. In cases where there is no bleeding the affection is called "blind piles." The tumors are also divided into internal, those which are within the anus, and external, those which are without the sphincter. Hemorrhoidal tumors differ in character; some are simple and consist of varicose veins in clusters, forming something of a tumor; these tumors are filled with liquid blood and can be easily squeezed out; the others are harder and contain coagulated blood; there is more or less inflammation present and consequently an exudation of lymph in the contiguous parts and a soft, spongy tumor is usually the result.

This derangement causes the animal great annoyance and at times more or less fever, loss of appetite and a falling off in flesh.

Treatment: The pig should get aloes, ten to fifteen grains, calomel, six to eight grains when this operates it will relieve the congestion and often cure. If not, open the tumor and squeeze out the blood and dress the part with a mixture of twenty grains of tannic acid, one ounce glycerine and one ounce water a little of this should be applied twice a day. Fifteen to twenty drops of turpentine at a dose twice a day is very useful balsam of copaiba in twenty to thirty drops once or twice a day is also good. Cold water injected several times a day relieves the inflammation, or a teaspoonful of cold water injected several times a day will be found beneficial. The animal should be fed on laxative food, compound liquorice powder is an excellent laxative in doses of thirty to forty grains once a day. There are a number of other remedies which are useful, but the above are among the best.

PROLAPSUS ANI.

This derangement is frequently met with in the pig. It is a protrusion of a part of the rectum caused by weakness of the muscles of the bowels and is often the result of either constipation or severe diarrhea and dysentery or whatever causes severe straining. It is usually the mucous membrane which protrudes. If it remains out for some time it swells and becomes of a dark color. Although it may have been protruded for several days it can be reduced and the animal recover.

Treatment: Bathe the protruded part for ten minutes with warm water to clean it, then bathe for ten minutes more with a mixture of two drams of acetate of lead, one ounce tincture of opium and one pint of water, then turn the pig on its back and push in the bowel, give two grains of opium to relieve the straining. It may be necessary in some cases to put in a stitch of catgut or silk thread across the opening to keep it from slipping out. The pig should have laxative food such as oatmeal gruel and if constipated give a dose of castor oil, this to be followed by from two to three drops of the fluid extract of nux vomica three times a day in the food. If it should be impossible to keep it in and the parts become mortified, remove it with a knife and if necessary stitch the bowel to the margin of the anus with catgut or silk some cases of this sort do well.

PERITONITIS (INFLAMMATION OF THE PERITONEUM.)

Peritonitis is an inflammation of the membrane lining the cavity of the abdomen, also reflected over the bowels.

Cause: Chills, operations, especially after the operation of castration, injuries such as kicks or blows from horses and the result of difficult parturition. The disease is an ordinary result of strangulation of the bowels. It is also caused by foreign bodies penetrating through the walls of the intestine

Peritonitis sometimes comes on in the course of other diseases, particularly hog cholera, which it brings to a rapid, fatal issue.

Symptoms: The disease is ushered in with a chill, there is great pain, the animal moves about in a very stiff manner and suffers intense pain, the abdomen is very tender to the touch and although it might be slight it will make the animal squeal. The muscles of the abdomen are contracted which gives the animal a tucked up appearance, there is usually some tympanitis which makes the muscles swell upwards and arches the back, the bowels are usually constipated and the urine is scanty and high colored, vomiting is very often present and it causes the animal great pain and increases the inflammation and should be prevented, the pulse is very frequent, small and wiry, from 110 to 140 or more per minute, the breathing is short and fast and there is a marked expression of distress characterized by the contraction of the muscles of the face. The cause of peritonitis is generally rapid, and in fatal cases the animal usually dies within twenty-four hours, but in some few cases it may run for one or even two weeks. When the fatal termination approaches, the pain suddenly subsides or ceases altogether and the animal may lie perfectly quiet; at this stage the pulse is very small or may be imperceptible, the legs are cold and the skin of the body is usually of a purple color, green or black matter will often be vomited, coma or convulsions take place and the animal soon dies.

If a favorable termination is indicated the pain will gradually subside, the muscles of the abdomen and face relax, the animal becomes brighter and takes notice of things around it, the pulse is less frequent and stronger, the breathing becomes more regular and the animal either lies quiet or may begin to move about looking for food. In the greater number of cases when they begin to improve they make a complete recovery.

Post mortem appearance: In cases which die early in the attack the only morbid appearance will be a dark redness with more or less swelling of the membrane and may be black or blue in spots. In cases of long duration there will always be more or less fibrous exudation upon the surface of the membrane and a considerable quantity of free liquid in the cavity. In some cases the fibrous exudation may become organized and join the other portion on the opposite side. The membrane may be somewhat of a dull white color, but in the majority of cases it will be found of yellowish turbid or dirty brown color and often milky, seropurulent or bloody. In some cases pus is found in pouches of the false membrane. The neighboring organs are more or less implicated.

Treatment: A dose of castor oil or sulphate of magnesium should be given to clean out the bowels. If the stomach will not retain the oil or salts give ten to twelve grains of calomel in pills the physic will act as a revulsant besides removing secretions and stimulating the portal system.

This should be followed by from two to three drops of the fluid extract of veratrum viride every hour until the pulse is reduced in force and frequency. If there is a tendency to vomit give one grain of opium. To relieve pain and keep up the action of the mercury give one or two grains of opium and two to four grains of calomel twice a day. If all medicine is rejected by the stomach give one to two drops of the wine of ipecacuanha every two hours and give injections of warm water to empty the bowel, then inject two drams of laudanum mixed in starch gruel repeat this every two hours. Mustard made up with boiling water and well rubbed into the skin of the abdomen or hot fomentations may be of use. If the disease does not yield to the above treatment in two or three days it will be necessary to have recourse to mercurial impression this is best done by giving from two to four grains of calomel combined with half a grain of opium every four hours until the animal is salivated. When the acute stage has passed the animal should have stimulants beat up eggs and put in a tablespoonful of whisky or brandy and give it frequently. If the pulse becomes very weak give aromatic spirits of ammonia in teaspoonful doses every two hours it should be taken in a little cold water. Later turpentine in doses of from ten to fifteen drops three times a day will be found very useful. Feed on well boiled oatmeal gruel cooled by good fresh milk.

Fig. 141.

Liver of a Hog—posterior view. a, Right external lobe; b, Right internal lobe; c, Left external lobe; d, Left internal lobe; e, Spigelian lobe; f, Posterior cava; g, Quadrate lobe; h, Gall bladder; i, Cystic duct; k, Ductus Choledochus.

CHAPTER VI.

THE LIVER OF THE PIG.

The liver of the pig consists of four chief lobes: a right and left external, a right and left internal, a small quadrate and a spigelian lobe. The gallbladder is large and is attached to the internal lobe by cellular tissue. Diseases of the liver are quite common in the pig, but they are so difficult to diagnose that they are only discovered on post mortem. Diseases of this organ are very often associated with other diseases. By careful observation and practice a few of them can be determined in the living pig.

HEPATITIS. (INFLAMMATION OF THE LIVER.)

Inflammation may affect the substance of the liver, its investing peritoneal membrane, or both, and may involve the whole organ or only a part of it.

Causes: Changes of temperature, such as from cold to heat, injuries, and the result of other diseases. It often occurs in high-bred pigs as a result of feeding with an excessive amount of stimulating food and want of exercise and is most commonly met with in hot weather.

Symptoms: The animal appears dull, refuses food, if made to move will go stiffly and may be lame in the right fore leg; there will be tenderness on pressure over the ribs on the right side which will not be the case if pressure should be applied to the left. There will sometimes be a yellowness of the visible mucous membrane and of the skin. The functions of the liver are arrested, thus the secretions of bile are not carried on and on this account the bowels become torpid, and the feces of a chocolate color. Sometimes they are affected by looseness and the feces are generally unhealthy, evinced by an excess or deficiency or perverted state of the bile. There is usually a cough which may arise from the pressure of the liver against the lungs or from sympathy. The urine is high colored and scanty, the respiration is somewhat impeded and is short and jerky, the pulse is soft weak and frequent. With these symptoms and the absence of other diseases we may conclude that we have a case of hepatitis. There is a chronic form of this disease which I have met with in pigs. The symptoms are a dry, scurfy skin, with an unthrifty appearance, in the majority of cases there will be a diffused yellowness of the mucous membrane; the animal falls off from condition and has a disinclination to move about; the pulse and respiration are unaffected; the feces are of a dry clay color and the urine is usually scanty and high colored of a deep yellow color. Hepatitis is so fruitful a source of other morbid affections that no time should be lost in its treatment.

Treatment: In acute hepatitis when the pulse is full and strong, the pig should get a dose of from one to two ounces epsom salts to act as a revulsant by depleting the system and indirectly relieving the portal veins. Mercurials are especially indicated in consequence of their property of increasing the hepatic secretions and thereby directly unloading the congested vessels of the liver. For this purpose give from five to fifteen grains of calomel, this should be followed by a dose of effusion of one ounce senna leaves; after the physic has operated freely give three grains of calomel morning and evening for a few days. If there is much pain add one grain of opium to the calomel; also give from ten to fifteen grains of nitrate of potassium twice a day to keep the kidneys active. In the chronic form give one to two ounces castor oil and after this has operated give from ten to fifteen drops of nitro-muriatic acid three times a day in a little water. To improve the appetite give a teaspoonful each of tincture of gentian and ginger and two grains of quinine at a dose in water three times a day. If the bowels are inclined to be constipated give ten grains of aloes at a dose at night. There are a number of other diseases of the liver which are not easily detected during the life of the animal, and may be taken for chronic hepatitis, such as hypertrophy; this is an overgrowth of the organ, and in some few cases can be diagnosed by a bulging of the right side of the animal. It may not affect the health of the animal for some time, but sooner or later the structure of the liver becomes impaired.

Atrophy. This is exactly the reverse of hypertrophy and very soon deranges the system, causing death.

Induration. The liver often acquires an increase in its density and hardness from depositions or new formations in its substance.

Softening. This is often the result of inflammatory action, but it may also occur without this cause.

Fatty liver. This is an adipose degeneration of the liver, which usually increases in size. The weight is not increased in a degree corresponding to the enlargement. The characteristic hue of fatty liver is a pale yellow or cream color, diversified by brownish, orange or reddish spots. It is softer than a healthy liver. When cut it has a brownish or pale yellow color, which is usually modified by innumerable red spots. It has a greasy feeling between the fingers.

Waxy liver. Pathologists now believe it to be the result of a peculiar degeneration or deposition having no resemblance whatever to fatty matter. The liver is much enlarged and at the same time denser than in health. It is usually of a pale or fawn color, but sometimes red from congestion. It is tough in texture, and when cut presents a uniform compact, smooth, somewhat shining or translucent surface, pale or yellow, and not unlike the rind of bacon or yellow wax. (Wood.)

Serous Cysts and Hydatids. This is a common disease in the pig as well as in the other domestic animals. I have seen a great many cases of it. The cysts consist of sacs containing watery fluid. There are two kinds, one consisting of a single sac, the other of a sac containing within it one or more sacs. There may be only a few cysts scattered over the liver, or in bad cases it may have the appearance of being converted into cysts, filling the whole liver. These cysts contain a parasite called a hydatid the origin of which is uncertain. When upon the surface of the liver they may break and cause peritonitis by the escape of their contents, producing death. An animal may be affected with this trouble for a long time and the health remain good, as I have found them in the liver of pigs that have been killed for food, and they did not seem to have affected them in any way. On the other hand I have examined pigs that had died, and found the liver in a morbid condition, having undergone a number of changes affecting the health of the animal sufficiently to cause death.

JAUNDICE.

This is not a common disease in the pig. It is characterized by a yellowish color of the skin, eyes and urine. It is the result of absorption of bile into the tissues of the body, and is usually easily remedied in the young animal.

Symptoms: There is usually loss of appetite, dulness, sometimes vomiting, and other signs of deranged digestion, soon the feces become of a clay color, and the membrane of the eyes becomes yellow, and if there is white hair the skin will be yellow, then it gradually becomes dark and remains in that condition for a few days or sometimes weeks, then it begins to disappear. The bowels are usually constipated and the urine of a deep yellow color. In the great majority of cases there is not much fever and the animal is usually well in two weeks.

Treatment: It is not easy to find the cause in the pig, but nearly all cases of jaundice require medicine to eliminate the bile from the system. It is therefore necessary to clean out the bowels, and this is best accomplished by giving from one to two ounces epsom salts; after this has operated give three grains of calomel at a dose three times a day for three days. Then give one ounce castor oil and half an ounce of turpentine, to be followed by ten drop doses of nitro-muriatic acid.

CHAPTER VII.

HERNIA. (RUPTURE.)

Hernia is the rupture of the walls of any organ, but is most frequently applied to a protrusion of the bowel through an opening, whether natural or artificial. Hernii are classified according to their position. The ones most commonly found in the pig are umbilical and scrotal hernia. The former is not often seen in the pig and is usually congenital and makes its appearance at birth or if not then, shortly after. It consists of a protrusion of omentum or intestine through the umbilicus, and is caused by a nonclosure of the navel opening.

Symptoms: There will be a bunch at the navel opening, soft to the touch and fluctuating, and if pressed upon will pass into the opening, to return as soon as the pressure is removed. It does not inconvenience the animal as long as it does not become strangulated, which it seldom does, and if not increasing in size it is better not to meddle with it, but if it is growing larger the sooner it is reduced the better. I have seen cases in the pig in which it became so large as to trail on the ground, and such cases cannot be successfully treated.

Treatment: Turn the animal on its back and press the intestine in; when this is done pass a skewer through the skin close to the walls of the abdomen; then tie a cord around the skin between the skewer and the walls of the abdomen; the cord should be tied tight enough to stop the circulation, but not tight enough to cut too soon into the skin. It will be better to tie on a second one on the third day than to tie the first one too tight. This method causes an outpouring of serum which fills up the opening, and in two weeks' time it becomes organized and the cure is complete.

SCROTAL HERNIA.

This form is where the intestines have passed into the scrotum or pouch through an opening or canal leading from the abdomen to the scrotum. In some cases in young animals this canal is large, thus allowing the intestines to escape through it into the pouch.

Symptoms: The scrotum will be larger than natural, and when pressed upon it will be soft and doughy.

Treatment: Place the animal on its back and press the intestines into the abdominal cavity, draw up the scrotum and testicles as far as possible, then put on a wooden clamp below the testicles, let the clamp remain on until it sloughs off and by that time the opening will be closed a plain clamp, no caustic is needed. If much swelling should take place bathe with hot water, and after each bathing use a little of the following lotion: acetate of lead, half an ounce, sulphate of zinc, half an ounce, water, one quart.

PLATE 52

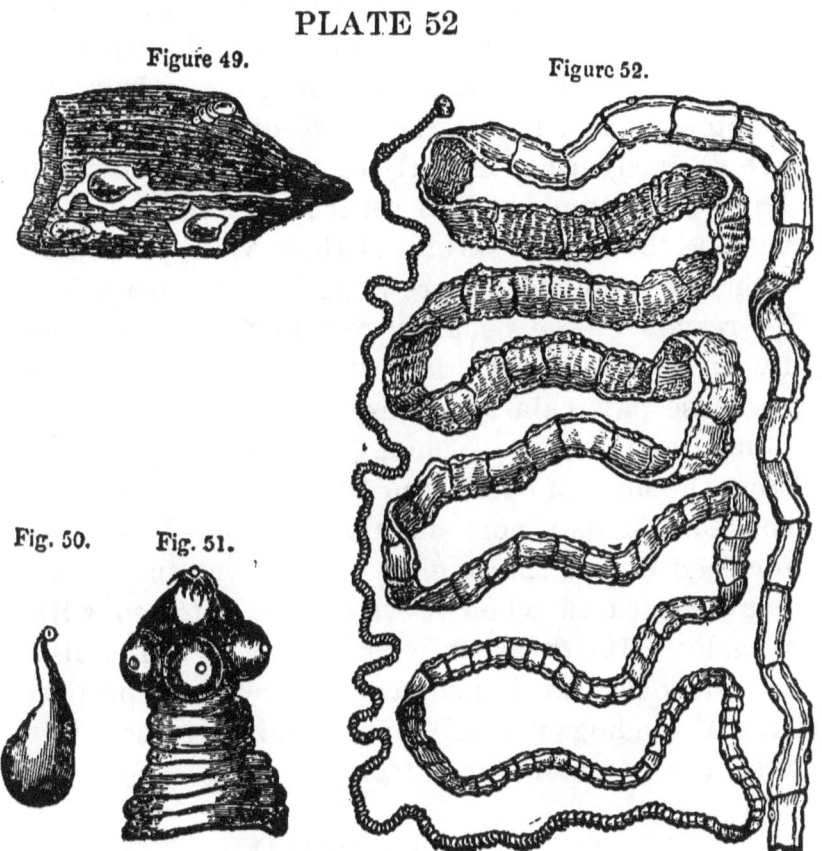

Figure 49.—Pork "measles;" natural size; Hearth and Home, after Owen.

Figure 50.—Young tape-worm from measles of pork; and Fig. 51, head of same, more enlarged; Hearth and Home, after Owen.

Figure 52.—Pork tape-worm (Tenia solium), less than natural size; Hearth and Home, after Owen.

THE INTERNAL PARASITES OF THE PIG.

The parasites of the pig are so intimately connected with those of the human being that there is no doubt but that in certain stages of their development nearly all the most dangerous kinds are derived, either directly or indirectly, from the hog to mankind, and in return man furnishes to the hog the eggs of several of their worst parasites. Some of these worms are kept in existence solely by reason of the peculiar relations existing between man and the domestic animals. This is the case particularly in respect to the most common, tape-worm, of men, derived respectively from the flesh of cattle and hogs when imperfectly cooked. Measly pork should be avoided as unfit for food unless thoroughly cooked, yet such pork has been eaten when it was badly infested with measles. It can be easily known by its spawn-like appearance. In this country a very large proportion of the hogs not only harbor this entozoon, but also the common lung-strongle.

GENERAL OBSERVATIONS.

Worms, sometimes on account of their movements and the interruption they offer to the contents of the bowels, and their other modes of irritation, very often cause uneasiness and pain. From sympathy there is very frequent sensation of itching at the anus and nose, producing a disposition to scratch or rub the root of the tail and poke the nose into the ground. This is a very common symptom of worms. The bowels are sometimes constipated and at others relaxed with straining and mucous or blood discharges; the mucus which is passed may be in shreds, which are sometimes mistaken for fragments of half digested worms. They also interfere with digestion, both of the stomach and intestines, which is indicated by undigested food in the feces. The appetite is very variable it may be natural in some cases and in others deficient, depraved or craving usually the animal has a ravenous appetite, eating material which it would not touch in health. Bleeding from the nose, cough and swelling of the lips are also symptoms. The effects of worms frequently extend beyond the alimentary canal. Among the most common of these affections are derangement of the nervous system, such as vertigo, chorea and fits, obstinate cough, swelling of glottis, dyspnea and palpitation. They also cause some fever and loss of flesh and several skin diseases. It is difficult to diagnose a case of worms, as the above symptoms may be caused from indigestion. A new means of detecting them is by the microscope; even if there are no fragments of the worms present in the feces their eggs can be detected. Suspected cases ought to be examined by the microscope, and if the examiner is acquainted with the character of the ova this will decide the case.

Causes of worms: There has been a great controversy on this subject, but no doubt the parasites obey the general law of nature in their development and growth. It is probable that the ova which are received into the alimentary canal are capable of development in a healthy state of this structure, as worms are sometimes found in the stomach and intestine of the hog. There are certain conditions of these organs which are favorable for the breeding of worms, such as indigestion, the accumulation of undigested food and of mucus and diseases of the intestine. The kind of food the animal eats in a raw condition containing the ova of worms makes the pig more liable to worms than other animals. All kinds of vegetable and animal food should be cooked: grains can be eaten raw with impunity; hogs should not be allowed to eat diseased meat unless it has been thoroughly cooked. I have known healthy pigs which were given the flesh of animals that died from disease to become badly affected with worms.

THORN-HEADED WORMS.

Figure 74.

Figure 74.—This worm is quite frequently found in the intestines of pigs; it is easily known by the peculiar proboscis which bears several circles of small but sharp hooks. They locate usually in the small intestines of the pig, where they fasten themselves by means of the spiny proboscis mentioned above, this being pushed into the lining membrane of the intestine; in some cases they bore through

this and migrate to other parts of the body, where their presence causes great disturbance. The eggs of this worm pass from the hog and are eaten by the grubs of certain large beetles; in the stomach of these grubs the eggs develop into embryos, or rather the embryos already developed are set free and bore through the intestine and locate themselves in the body of the grub. Here they become encysted and remain dormant until the grub is eaten by the pig and then once in the stomach or intestine of this animal it develops into a worm at once. The color of this worm is white or bluish white, the female being from five to twenty inches in length, while the male is from three to five inches long. The female is very prolific, producing immense numbers of eggs, which are of a somewhat oblong-oval shape.

Symptoms: Pigs may have a number of these worms without their causing any noticeable derangement; on the other hand they often derange the digestion and assimilation, causing loss of flesh and a general unhealthy appearance. The animal is usually hungry and may eat large quantities of food and yet remain thin. When a pig has such symptoms with the absence of any other ailment we may suspect that worms are the cause. In very bad cases the pig becomes weak in the loins and the membrane in the corners of the eyes swollen, red and watery; the animal suffers pain, which is indicated by it continually grunting or squealing; such hogs are usually bad tempered and will bite and snarl at the other pigs. In some cases the weakness increases and the animal is unable to stand and soon dies.

Treatment: This worm is easily removed by medicine. I have had good results from the following: Give half an ounce of the fluid extract of spigelia and senna at a dose every four hours until purging takes place, or thirty grains of koosin as a pill; one dose of this is usually sufficient. Another good remedy is to beat up two ounces of pumpkin seeds into a pulp with sugar and give at one dose; this should be followed in four hours with a brisk physic, castor oil or epsom salts. Santonine is also useful in from three to five grain doses made into a pill. Chenopodie oleum (worm seed oil) in from twenty to thirty drops may be given in a little syrup, followed in two hours by a purge. These medicines should be given on an empty stomach. If the animal has become very weak the strength should be kept up by stimulants such as small doses of whisky and cod-liver oil, or whisky and eggs; and if there is fever two to four grains of quinine should be given at a dose three times a day.

OXURIS VERMICULARIS (PIN WORM.)
Figure 78.

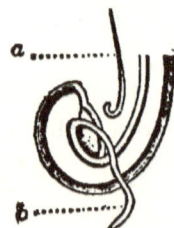

Figure 78.—The seat of this worm is the rectum,

but they are sometimes found in the colon and have been seen in the stomach; on this account they have been called the maw worm. They are usually about half an inch long and white in color; they multiply very rapidly; their eggs are very small and are often deposited on the grass and may be washed into the streams or ponds of water and in this way enter the stomach and bowels. They are found in all the higher animals.

Symptoms: They usually cause itching at the root of the tail or by reflex action cause derangement of other parts of the body. Though productive of great annoyance and even suffering, they do not usually injure the health of the animal.

Treatment: Clean out the rectum by injections of warm water. Infuse two ounces of quassia chips in a pint of boiling water and when cool inject it into the rectum; repeat in a week if necessary. A brisk purge will often wash out a number of them.

TRECOCEPHALUS DISPAR (LONG THREAD WORM.)

This worm is found most frequently in some part of the colon and cecum, but sometimes in the small intestines, either loose or with its anterior capillary portion inserted into the mucous membrane. It is often observed in great numbers in animals that have died from some acute disease. I have found numbers of these worms partially buried in the mucous membrane, but they did not seem to have caused much disturbance. They are about half an inch to one inch and a half long and about as thick as a common thread, and are very active in their motion. From experiments that have been made it appears that the ova are never developed in the animal body, but being discharged with the feces retain their vitality for a long time, and if placed in water become at the end of about eight months and a half developed into embryos, about one three-hundredth of an inch in length. It is probable that these are carried by the rain and other means into streams, wells, etc., whence the drinking water is derived and thus they become fully developed. There are no particular symptoms by which we can detect this worm from others during the life of the animal, and the treatment would be the same as for the thorn-headed worm. Very often when worm medicine is given a variety of worms will be discharged at the same time.

ASCARIS SUILLA (ROUND WORM.)

This is a round, smooth worm of considerable size; the male usually reaches when fully developed six inches, and the female may be twice that length. This worm usually does no harm when there are only a few present. The animal will likely keep in good health, but when they are numerous they will disturb digestion and cause colicky pains, loss of flesh, dry hair, morbid appetite, restlessness and nervous twitching, and in some cases fits. Cases are recorded in which they worked through the walls of the intestines and, reaching some of the other organs of the body caused death. This worm generally inhabits the small intestine, but not infrequently finds its way forward to the stomach or backwards to the rectum and sometimes escapes from the intestine through the anus. This worm has also been found in the biliary duct, gall bladder and the substance of the liver. There has been a number of experiments made to find the origin of this worm, and it is found that the eggs of this worm are passed from the bowel. They retain their vitality for a long time; they appear never to be developed in the bowels, but when discharged and kept in water they begin to show signs of life and in about seven months contain embryo worms one one hundred and twentieth of an inch in length. These have not been seen to break shell but the ova carried into streams, ponds and wells sometimes probably find an entrance into the stomach with the drinking water, when the embryo escapes from its shell and completes its growth in the intestine.

Treament: The best remedy for this worm is the fluid extract of spigelia and senna given in half ounce doses every four hours until it causes purging. Worm seed oil (chenopodium) in doses of from five to ten drops given in a tablespoonful of castor oil is also good. Turpentine in doses of from fifteen to twenty drops three times a day followed by castor oil or epsom salts is useful. The cedar apple, an excrescence found on the red cedar has been used with good results in doses of from twenty to twenty-five grains of the powder, repeated three times a day, followed by a physic.

SPIROPTERA STRONGYLINE—RUD.

There are a number of small, whitish or reddish round worms which taper somewhat towards the anterior end, or towards both ends. The head is small with small papillae or naked; the male grows to about half an inch long or more; the female one-third of an inch or more; it lives in the stomach of the pig, but generally does not produce any serious disease. The fluid extract of spigelia and senna in half ounce doses given every four hours until purging ensues usually dislodges them.

SCLEROSTOTUM DENTATUM (DIESING.)

This is a small worm living in the intestines of swine. The male is about one third of an inch long; the female half an inch long; the body is of a dark color and the surface is finely marked with transverse striae. It is quite slender and tapering at each end, but the male has near the tail a three-lobed expansion. The eggs are laid in the intestines, from which they pass out into the open air and are soon hatched. The mouth of this worm is circular and armed with six teeth, by means of which it attaches itself to the intestines and pierces the tissue, feeding upon the blood. If there are many of them they create such a drain on the system of their host as to weaken and possibly destroy it. It may also by its irritation of the bowels cause serious trouble and disease. An active purge is the best remedy for the removal of this worm.

STRONGYLUS DENTATUS (RUD.)

Figure 80.

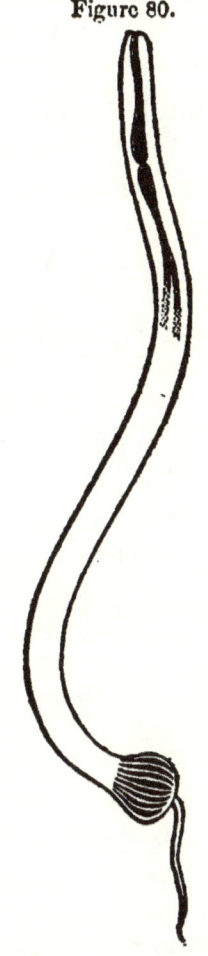

Figure 80.—This worm is found in the intestine of the hog. It is a slender filiform worm about half an inch long; the head is obtuse and surrounded by six acute papillae; the oesophagus is short, thick and muscular; in the male the tail is

truncated and provided with an oblique bursa; in the female it is elongated and slender, ending in a fine point; the genital opening is near the posterior end. The history of this worm is not known. It does not seem to do much harm. The usual treatment for worms is nearly always effectual in bringing them away. I have seen quite a number of them mixed with other worms in the feces of a hog that had been treated for worms.

STRONGYLUS ELONGATUS.

This species live in the lung and air passages of the pig. This worm is about one to one and one-half inches long. They often occur singly or several together. When they are numerous they set up great disturbance, often resulting in the death of the host. The first symptom of the disease is a cough, usually slight at first, but soon becoming very distressing, and the pig shows signs of suffocation, which sometimes takes place, or inflammation may set in and carry the animal off. This disease is often taken for catarrh or some other respiratory trouble and it is very difficult to diagnose unless a worm which has been coughed up may be seen protruding from the nose.

Treatment: This is not easy in the pig. Small quantities of turpentine injected into the nostrils may reach the worms. Turpentine given in teaspoonful doses three times daily will sometimes be of use, as the turpentine is partly eliminated by the lungs. The inhalation of the fumes of carbolic acid is also useful.

KIDNEY WORM (EUSTRONGYLUS GIGAS.)

Figure 81.

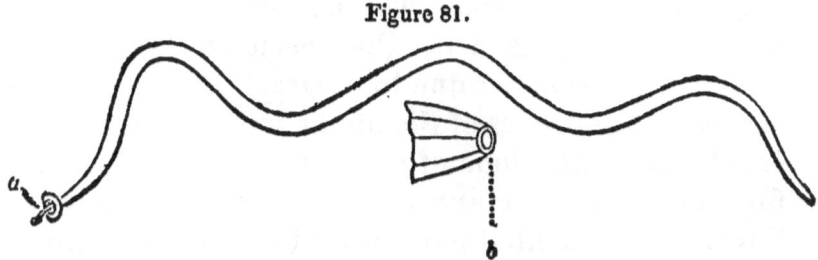

Fig. 81.—This worm is found in the kidneys of all the domestic animals and in man, although it is very rare. It is a large worm and it is said that sometimes the female becomes three feet long and half an inch in diameter, although usually much less. The male becomes ten to twelve inches long. I have never seen any of them so large, as they are usually only a few inches long. The body is smooth, round and tapering somewhat to each end, and of a deep red color. When such worms are present in the kidneys they gradually but completely destroy the substance of the kidney which becomes filled with purulent matter, upon which the worm feeds, while the walls often become hardened with calcarous deposits. The effects and symptoms are the same as in other acute diseases or abscesses in one of the kidneys. The only positive proof of the presence of the worms would be the discovery of the eggs in the urine. It is probable that no remedy can be applied when the parasite is once lodged in the kidney. The history of this worm is not fully known. (Verill.)

I have found live worms in the kidneys of both the pig and dog and the kidneys were perfectly healthy and neither animal seemed to be in any way affected by them. The loss of power of the hind parts of pigs which has been attributed to kidney worms, is not due to a parasite, but to paralysis of the muscles of the hind parts. I have made careful investigations of such cases, but failed to find any worms or any disease of the kidneys. Paralysis of the hind parts would not be the symptom of kidney disease.

TRICHINA SPIRALIS.

Figure 76.

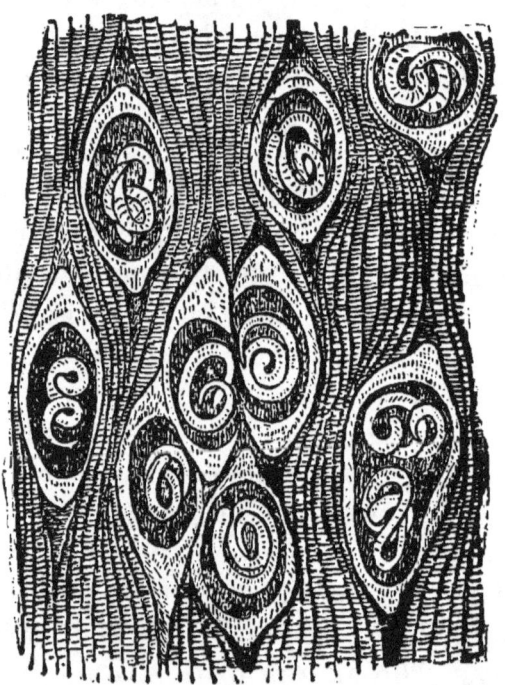

Figure 76.—A small piece of human muscle containing encysted young of Trichina spiralis Owen, enlarged forty-five diameters. From Hearth and Home, after Leuckart.

Figure 76.—This very minute worm is found in the larva stage in large numbers in the flesh of some pigs, dogs, cats, rats, mice, rabbits, guinea-pigs and many other animals, in the natural state in the intestines of the same animals. The male is very small, measuring only one-eighteenth of an inch. The female is stouter and longer than the male, measuring about one eighth of an inch. The young trichinae occur embedded in the muscles of the pig and various other animals. They are so small as to be quite invisible to the eye and millions of them may be in the flesh of the pig without producing any unusual appearance; even an expert could not detect them without the aid of the microscope. This is why so many deaths occur from eating pork filled with this parasite. When these little worms are first introduced into the muscles of the pig they are free and coiled up among the fibers of the muscles; but after a few weeks they become incased in minute whitish, elongated cysts, supposed to be the result of irritation set up by their movement and feeding. Figure 76. In a year's time these cysts become calcified by a deposit of carbonate of lime in the membrane; when this takes place minute white specks about the size of hemp-seeds may be seen in the muscles. When the worms are inclosed in the cyst they lie dormant and although they may live for years and even weeks after the death of their host they can do no further harm, unless they are eaten by man or some animal. Each cyst contains a little slender worm about one twenty-fifth of an inch long and one seven-hundredth of an inch in thickness coiled up in two or three turns; the size of the cyst is about one-eighth of an inch long and one one-hundred and thirtieth thick. If the flesh of the pig containing these worms be eaten by man, they become liberated in the stomach and, entering the intestine, attach themselves to its soft lining, and there, surrounded with abundance of food, they grow very rapidly and become mature, with fully developed sexual organs, in two days. The females are more numerous than the male and become about one eighth of an inch long when full grown. They pair as soon as mature and the male soon dies, but the female begins to give birth to living worms in five or six days from the time it enters the stomach and lives long enough to produce a brood of from five hundred to one thousand young worms each. As one ounce of pork often contains a quarter of a million or more of the worms, it is not surprising that the million of adult worms and their offspring sometimes resulting from a single meal of raw pork should by their presence, produce great irritation and inflammation of the intestine and violent diarrhea and vomiting, which are often the first symptoms in severe cases. But the young worms as soon as they are born, begin to eat or force their way through the membrane of the intestine into the minute blood vessels and other organs, thus vastly increasing the irritation. Entering the circulation they are carried by the blood to the heart, thence to the lungs and then become diffused through the whole system. (Verrill.) Some other observers think the young worms force their way directly through the intestine and all the intervening organs, until they reach a suitable habitation in the voluntary muscles. According to Dr. Leuckart they travel by way of the inter-muscular connective tissue, and are found most abundantly in the groups of muscles nearest the abdominal cavity, especially in those that are smallest and have the most connective tissue. The cysts containing trichinae were first observed in human muscles in 1882, but the worms were first named and described by Owen in 1835, but were only regarded as anatomical curiosities of no practical importance, until 1860, when Zeuker proved that they are capable of producing the severe and often fatal disease now well known under the name of Trichiniasis, but which has been previously confounded with typhoid fever, inflammatory rheumatism or rheumatic fever, poisoning and other diseases.

CHAPTER IX.

DISEASES OF THE RESPIRATORY ORGANS.

Inflammation may occupy a distinct portion of the respiratory passages and terminate where it began, or it may affect several portions successively or all at the same time, forming one continuous disease. It will be more convenient to consider each set separately. The pig is a very bad subject to examine as it is impossible to keep it quiet, and if we try to do so we excite the animal so much that both the respiration and the circulation are much increased in frequency, therefore we cannot get much aid in this line.

INFLAMMATION OF THE NOSTRILS OR CORYZA.

The same mucous membrane lines the nostrils and the sinuses of the head and face and these parts are all liable to become inflamed at the same time, constituting what is called a cold in the head, and very often the membrane of the eyelids will be affected at the same time through sympathy.

Causes.—The most frequent exciting cause is exposure to cold such as lying in cold, damp places in cold weather, especially cold east winds and rain. Some pigs are much more susceptible to colds than others. This disease is often epizootic, affecting nearly all the pigs in a neighborhood at or about the same time.

Symptoms.—The first symptom is that of dryness with some swelling of the membrane, and irritation which causes sneezing; this is generally followed by copious discharge of a thin, acrid fluid, which irritates the margin of the nose and the membrane, increasing the inflammation. The nostrils are partially closed by the swelling, which causes the animal to make a snuffling sound; the eyes become red and watery, either from sympathy or by the continuous extension of the inflammation up the lachrymal passages. In the majority of cases there are no constitutional symptoms; the pulse remains natural and the appetite is unimpaired. In some few cases there will be more or less fever indicated by loss of appetite, hot, dry skin and a somewhat excited pulse. The complaint usually attains its height in three or four days, then it begins to abate. The secretions from the nose are thicker and less copious and sometimes assume a yellowish color. If there has been any fever it gradually subsides and recovery is usually complete in from seven to ten days. If it should extend to the throat and downwards it would constitute laryngitis and bronchitis, which will be described under their respective heads.

Treatment: Most cases of this disease are so mild that no medicine is needed. The pig should be kept in a comfortable place at night and allowed to run at large during the day. A warm mash at night with a little good ginger in it will be all that is necessary. If there should be fever give the animal a dose of epsom salts and follow this by giving small doses of nitrate of potassium, ten to fifteen grains. If the nostrils should get very much filled up put a piece of camphor in hot water and hold it under the nose for ten minutes; or take a small bottle of tincture of iodine and hold it under the nose; the heat of the hands will cause it to give off vapor of iodine. A few doses of quinine often affords relief.

CHRONIC INFLAMMATION OF THE NOSTRILS OR OZENA.

When simple catarrh continues beyond the usual period it is apt to become chronic. The mucous surfaces have become weakened or ulcerated, giving rise to a yellowish muco-purulent discharge, or mucous membrane may become thickened and make breathing somewhat difficult. I have known young pigs to die of this disease.

Treatment: Give the pig from eight to ten grains of sulphate of iron in its food twice a day. In young pigs when they get badly stuffed up the nostrils should be cleaned out and steamed with hot water with a little carbolic acid in it. In some cases it will be necessary to inject into the nostrils a solution of sulphate of zinc, beginning with four grains to the ounce and rapidly increasing it to twenty grains to the fluid ounce repeat several times a day. Fowler's solution of arsenic in from three to five drops given in the food three times a day is useful; this can be continued for a month if necessary.

MALIGNANT CATARRH.

This disease in the pig somewhat resembles glanders in the horse. It is not common in this country, and when it advances to the second stage it is better to destroy the animal and thus prevent any danger of it spreading to other hogs, as it is useless to try treating such a case.

Symptoms: There is considerable discharge from the nostrils and eyes. The eyelids are swollen and the animal keeps in dark places, as the light seems to annoy it. The disease extends gradually to the back part of the mouth and throat, causing swelling of the mucous membrane resulting in suffocating cough with difficult breathing. The nose becomes thick and ill shaped; the discharge becomes very offensive and often mixed with blood; the animal has considerable fever; the breathing is very frequent, the pulse is rapid and the heart feeble; there is usually great thirst; the animal refuses all food and loses flesh rapidly; the feces are black and the urine high colored, and in this stage if it is not checked the pig soon dies from weakness and suffocation.

Treatment: Give two ounces of castor oil and follow this by giving from five to eight drops of nitrohydrochloric acid at a dose three times a day in a little oatmeal gruel. If the animal is weak give from two to four grains of quinine, in two tablespoonfuls of whisky. The head and face should be bathed with acetate of lead half an ounce to the quart of water. If the cough is troublesome give from fifteen to twenty drops of tincture of opium in a spoonful of water. When there is a tendency to suffocation gargle the throat with tincture chloride of iron, a teaspoonful to the ounce of water. Bathe the neck with hot water and then rub on mustard. Steaming the nostrils in this complaint is useful. If the animal does not improve in a few days it will be better to destroy it.

QUINSY OF THE PIG.

This disease is characterized by sore throat and the appearance of a swelling on the sides of the neck at the angle of the lower jaw and sometimes extending between its wings. When the swelling is large it presses on the larynx (upper part of the windpipe), causing difficult breathing, and in some cases suffocation. This is a local disease and one common to the pig, and usually yields to treatment.

Treatment: Give the pig all the cold water it will drink. Put half a dram of chloride of ammonia in half a pint of water and if the animal is thirsty and will drink water, put it in the drinking water; if not, give it with a bottle, repeat three times a day. Bathe the neck with hot water and rub on a little camphorated liniment. If the breathing is difficult give three drops of the fluid extract of belladonna and ten grains of chlorate of potassium at a dose three times a day; steam the nostrils three times a day with hot water, with a piece of camphor in it. Binding hot cloths around the throat is useful.

LARYNGITIS AND PHARYNGITIS (SORE THROAT.)

This is a very common disease in the pig, and is characterized by cough, difficulty in swallowing liquids, and rough breathing.

Causes: The causes of sore throat are changes from heat to cold or from a cold place to one that is hot and badly ventilated, lying in cold, wet straw and changes in the atmosphere. Pigs that have been driven a distance and are over-heated and cool off suddenly are liable to get sore throat.

Symptoms: There will be a hard, dry cough, difficulty in swallowing food and water, impaired appetite, froth from the mouth and more or less roughness in the breathing; the cough is often spasmodic and distressing; there is no external swelling and it is easily distinguished from quinsy on this account. In bad cases there will be considerable fever with a fast full pulse and some elevation of the temperature; the skin will be hot and dry and the hair rough; the pig will lie most of the time unless the cough is very severe, when it will stand up until the fit of coughing passes. This disease is usually manageable, though in a few instances, when very intense or attended with an unusual degree of serous effusion into the submucous tissue, it becomes very alarming and even causes death. Its great danger under these circumstances is owing to the narrowness of the passage through the chink of the glottis, which is closed by its walls being swollen, probably being aided by spasms of the muscles, thus preventing the admission of air so that the pig dies of true asphyxia. There is no other portion of the respiratory passages in which an equal extent of inflammation is capable of producing the same fatal results. I have made postmortem examination of young pigs which died of this disease, where the glottis was completely closed, all the other organs being in perfect health. Death is usually very rapid in such cases. I have known several instances in which the animal died in ten hours from the beginning of the attack.

Treatment: Give from one to two ounces of castor oil; follow this by giving ten grains chloride of potassium and from two to three drops of the fluid extract of belladonna three or four times a day. Steam the nostrils with hot water and camphor every two hours if there is much difficulty in breathing. In some cases great benefit will be derived from giving an emetic. Mix four grains of potassium-tartrate of antimony and six grains of ipecacuanha in a little gruel; repeat in half an hour if vomiting does not take place; this will remove any collection of mucus in the throat and will also reduce the fever; this should always be done when there is any tendency to suffocation. Half a teaspoonful of the compound syrup of squills is often useful repeated several times a day. If the cough is very troublesome give from fifteen to twenty-five drops of tincture of opium in a little water.

When there is much fever advantage can sometimes be had by giving two grains of calomel every two hours until it has made an impression on the pulse, or from three to five drops of the fluid extract of veratrum viride four or five times a day in a little water. Sometimes advantage is obtained from gargling the throat with alum and water, a teaspoonful of alum to a glass of water; this can be repeated every hour or two. The operation of tracheotomy cannot be resorted to with success on account of the shortness of the neck and the accumulation of fat.

CHRONIC LARYNGITIS.

This form is not common in the pig. The mucous membrane though inflamed maintains its integrity. There is little inconvenience; it is only when ulceration takes place that it injures the animal. A pig may be affected with this trouble and keep in good condition and the only thing noticeable will be a dry cough, which does not yield easily to treatment. The best remedy is the compound syrup of squills in thirty drop doses three times a day. If ulceration should take place there will be a little blood mixed with the mucus coughed up. This form of disease cannot be successfully treated in the pig. Giving turpentine in dram doses three times a day and making the animal inhale the fumes from burning tar will sometimes be beneficial and might be tried.

NERVOUS COUGH.

We meet with pigs having a cough, especially among the young ones; it is not attended with any symptoms of catarrh; there is no evidence of inflammation or irritation of any part of the respiratory passages, nor are there any of the abnormal sounds heard of a moist or dry character that can be detected. The trouble is purely a nervous one. It is caused by some morbid condition of the nerves of respiration or of the centers. The cough is usually dry unless the paroxysms are severe, then a little mucus may come up. It may last for months or it may pass off in a few weeks. It does not as a general thing disturb the health of the animal and resembles a mild attack of whooping cough in the human. It usually yields to treatment. Tincture of asafetida in dram doses three times a day or better if it can be obtained is allium or English garlic; the oil of garlic is the most convenient, dose from ten to fifteen drops three times a day given in a little syrup or dropped on sugar. I have found from experience that this form of cough will run its course without any treatment.

INFLAMMATION OF THE BRONCHIAL TUBES. (BRONCHITIS.)

Under this head we shall take in inflammation of trachea as well as the bronchial tubes. Bronchitis is not very common in the pig, although I have seen some well marked cases of it, especially in shoats of from six weeks to three months old, and it often proves fatal. It varies very greatly in degree and character. The inflammation very frequently begins at the nostrils, fauces or larynx and passes down the trachea into the bronchial tubes.

Causes: Cold in some form is the usual cause and is most common in the fall. Young pigs which have to lie out at nights exposed to cold rains or those kept in cold, wet houses are very apt to take inflammation of some part of the respiratory passages. Pigs should have a good, dry, warm place to sleep in at night. It is said that an excess of ozone in the atmosphere may occasionally cause it as well as coryza and laryngitis; this may be true as we sometimes find a number of animals affected with bronchitis at the same time.

Symptoms: There is a cough and usually some hoarseness with distressed breathing; the animal is restless, holding up its head as if it was suffering snuffing the air; there is fever indicated by dryness of the skin and fast pulse; the cough is at first dry and painful; as the disease advances the cough becomes softer and after severe coughing there will be more or less mucus of a white frothy color coming from the mouth; the appetite is impaired according to the severity of the fever. The disease generally lasts from four to ten days; at that time if the appetite is improving, the skin becoming moist and the cough less frequent, the animal makes a rapid recovery. In bad cases the inflammation sinks deeper into the small bronchial bronchi; the air enters with difficulty through the constricted tubes, causing great oppression and difficult breathing, and the animal suffers from a feeling of suffocation. Another danger arises from the small tubes becoming filled with mucus interfering with the due aeration of the blood. This blood, passing through the vessels of the brain, has a very depressing effect on the system and sometimes causes sudden death, or death may be preceded by coma or delirium. In very distressing cases there will often be a little blood mixed with the mucus. This is a much more dangerous disease in animals, especially the very young ones, as they seem to be unable to throw out the mucus.

Treatment of Bronchitis: In the early stages of the disease give one ounce of epsom salts to clean out the stomach and bowels and also to act as a revulsent; follow this by taking one pint of linseed tea with two drams of antimonial wine in it; give this in divided doses during the day. Other demulcents such as gum arabic in proportion of one ounce to the pint of water or effusion of slippery elm would be useful. If there is much cough give from one to two grains of opium in a little of the above emulsion; if there is high fever give three or four drops of tincture of aconite and from ten to fifteen grains of nitrate of potassium. In severe cases if the hog is strong a full dose of calomel, eight to ten grains, given at once, is very useful. We cannot with any satisfaction to ourselves or benefit to the pig use hot water to the surface of the body; but I think some benefit may be derived by rubbing the sides well with compound soap liniment or mustard several times. After the acute stage is passed the compound syrup of squills in doses of from twenty to thirty drops three times a day will be found serviceable. At this stage of the disease opium should be avoided as it is apt to arrest the secretions or prevent the expulsion of mucus from the small bronchi and therefore increase the distress and danger. It is better in the case of the pig not to restrain the cough, as it is often an effort of nature to expel the accumulated mucus. If the cough should be very distressing recourse may then sometimes be had to the fluid extract of hyoscyamus three to five drops at a dose every two hours or hydrocyanic acid in doses from five to fifteen drops every two hours. In the advanced stages when it appears to be verging on a chronic form, take one ounce of the bruised roots of senega and licorice, boil this in one and one half pints of water down to a pint, and when cool add one grain tartar emetic and two ounces of sugar, and give the animal a tablespoonful of this every two hours. If the pig is weak, give it five to ten grains of carbonate of ammonia in a little cold water every two hours. The animal's strength should be supported by good milk or eggs beaten up and a little whisky added to it. Keep the animal as comfortable as possible.

Post Mortem Appearances.—The bronchial mucous membrane is reddened, thickened, sometimes softened; in some cases there is ulceration and gangrene. Occasionally the redness is diffused, but more frequently in patches. In some few cases there will be abrasion. The tubes contain mucus in various states, sometimes blood and not infrequently pus. In cases where the animal has died from an accumulation of mucus in the tubes the lungs refuse to collapse upon the admission of air into the pleural cavities. Under the microscope the surface of the membrane may sometimes be seen deprived of its epithelium and the fibrous layer covered with a fibrous exudation in its place. The substance of the lungs is more or less affected, often they are congested in patches, sometimes there will be small abscesses filled with pus bearing a resemblance to tubercles when cut into; these are lined with false membrane and communicate with the bronchial tubes. Chronic bronchitis is not a usual result of the acute in animals; but if a cough with more or less discharge of mucus should remain after all acute symptoms have passed it may be called chronic and is best treated by tonics and stimulants, such as iron, quinine and whisky, or from five to eight drops of Fowler's solution of arsenic, three times a day in food. A very good one is ten grains of sulphate of copper in the food three times a day. Two drams of tar in the food twice a day I have found very effectual.

CONGESTION OF THE LUNGS.

This is a common disease in the pig, and consists of engorgement of the vessels of the lungs with blood and a detention of it in the capillaries causing what is known as pulmonary apoplexy. This state of engorgement is recognized as acute, passive and mechanical. The former is the one which we most frequently encounter and is readily recognized in the pig. When pigs are fat they have very little breathing space left and therefore are not in a fit condition for fast movement. When it becomes necessary to drive pigs in this state, particularly if the day is hot, great care should be taken not to push them, but allow them to rest occasionally. I have known pigs which have been driven or chased when they would stray from the herd to fall down and never rise again. The exertion caused the heart to propel more blood to the lungs than they could receive in a natural state, causing distention of the vessels which pressed on the air cells preventing access of air resulting in suffocation. When the animal shows symptoms of fatigue it should be rested for a short time and be given a drink of water. It is also caused by colds, badly ventilated houses and is the result of other diseases.

Symptoms: The animal is in great distress; if it is standing its head will hang down and its forelegs will be wide apart and its flanks heaving at the rate of one hundred per minute; the mouth will be partially open; the eyes are bloodshot; the heart beats tumultuously. In some cases there will be blood oozing from the nose; the pig soon becomes unable to stand and it lies down on its breast with its nose resting on the ground; the legs and ears are cold, and the mouth hot. If the animal is made to rise it will often squeal, but in the majority of cases it will not get up; the pulse is small and indistinct and the beat is difficult to make out; (which may be as many as one hundred and fifty per minute); the heart's action is rapid, jerky, disturbed, and tumultuous, but takes strength; auscultation of the chest is not satisfactorily made in the pig; but in some cases where the animal will lie quiet a minute crepitation or a fine sharp crackling sound will be heard; this sound may be diffused or it may only be detected in portions of the lung, but usually no distinct sound will be heard. Congestion of the lungs occurring as a result or in connection with other diseases, although the symptoms are not to all appearances so severe, are generally more fatal than when the result of over exertion.

Terminations: The great majority of cases which are severe and acute unless prompt treatment is had recourse to will terminate fatally. The condition of the congested vessels rapidly causes death from asphyxia. Milder cases may terminate in inflammation of the lungs.

Post Mortem Appearance: In pigs which die of this disease the lung tissue will be found greatly distended with dark colored fluid blood with occasionally circumscribed effusion of blood from rupture of minute and capillary vessels. The lungs are swollen and of a darker color than natural and their crepitant character is much diminished, although not altogether gone. Their normal elasticity and spongy texture is somewhat destroyed and although heavier than natural, still they will float on water. When the surface is cut there will be an oozing of blood and if squeezed it will drip as if from squeezing a sponge, and it is rendered frothy from the entanglement of air. The lining membrane through the entire bronchial tubes is reddened and covered with frothy mucus. In no part do we find effusion, save of serous material, the characteristic plastic exudation of inflammatory action being as yet undeveloped. Both sides of the heart, but particularly the right, with the large blood vessels proceeding to and from it, are more or less full of dark-colored blood, but not coagulated.

Treatment: As soon as possible give the animal a stimulant such as one dram of aromatic spirits of ammonia and one dram spirits nitrous ether at a dose in a little cold water, and repeat in half an hour if necessary; or give from one to two tablespoonfuls of good brandy or whisky; cover the body with warm blankets and rub the legs with a little of the compound soap liniment. If the animal improves keep up the stimulants for a day or two, but give them less frequently. Brandy or whisky will be found more useful in the later stage than the ammonia or ether. When relieved feed the pig milk and oatmeal gruel for a few days.

PNEUMONIA (INFLAMMATION OF THE LUNGS.)

Pneumonia may be defined as an inflammation of the true lung substance and the connective tissue. This disease attacks all kinds of animals, but there are none in which it is so fatal as the hog. The reason of this is in a great measure due to not detecting the disease early and after it has been detected the great difficulty we have in treating it properly in the pig.

Causes: Season and locality; it is more prevalent in the cold weather than in summer; sudden variations of temperature, especially if it is wet; particular states of the health, it being more liable to attack animals weakened by previous disease. It is sometimes the result of catarrh and sore throat especially if the animal is exposed to cold and wet while suffering from these diseases. It is caused by direct irritation, such as medicinal liquids and gases which find their way into the bronchial tubes and thence into the tissue; from wounds or portions of fractured ribs penetrating the substance of the lungs; it is also the result of foul air and is sometimes caused from the so-called epizootic influences or agents and from pulmonary congestion.

Symptoms: Pneumonia is usually ushered in by a chill, followed by fever, difficult short breathing, cough and, if the animal is forced to move, will show signs of pain and in some cases will squeal. Occasionally the symptoms of the disease are preceded by general uneasiness, dulness, loss of appetite, and more or less fever, when, as the disease advances, the pulse will increase in frequency and will often reach from one hundred and twenty to one hundred and fifty per minute. The cough is often very troublesome, dry at first, but as the disease goes on it will become softer. It is a cough which brings no relief to the animal; but when the mucus begins to form and the animal can bring some up it then gets relief. About this period if the fever abates the animal moves about and takes a little food, the pulse is less frequent, the breathing slower and easier, the chances are good for its recovery. On the other hand, if the fever persists and the pig becomes more restless and the advancing temperature which in the early stages of the disease would be one hundred and two to one hundred and three now jumps up to one hundred and five to one hundred and six, and the matter the animal coughs up is of a red, rusty color, the eyes sunken and the animal very weak, it will likely die. The duration of the disease is from eight to fourteen days.

Post Mortem. The general pathological condition of the lung tissue is hyperemia and swelling with a variable constituted fibrinous exudation, chiefly of the small bronchi and air cells, with usually a considerable quantity of serum surrounding the outside of the lungs. This is the usual appearance of those which die about the sixth or eighth day. The second stage, that of exudation, red hepatization or red softening, is characterized by a deep red, reddish brown or grayish-red color, the absence of crepitation under pressure. The diseased lung is so much increased in density that it will sink in water. The grayish color sometimes observed is owing to an intermixture of particles of black pulmonary matter and to the lighter hue of the interlobular tissue, which is occasionally less congested than the other parts. In some cases the softening is so great that it may be easily torn and the fingers may pass through its structure with very little resistance. When cut into with a knife it somewhat resembles the liver; on this account it is termed hepatized. When pressed between the fingers a reddish fluid will ooze out. The cut surface of the lung exhibits numberless minute granules which are probably the air cells filled with a concrete fibrous exudation. In red hepatization the bronchial tubes, the blood vessels, and the interlobular areolar tissue are still obvious to examination. The lungs do not collapse upon exposure to the atmosphere, as they do in health. In the third stage there are two conditions, namely gray hepatization and the other abscess. In the former the lung is compact and of a gray color, both externally and internally, and when cut into a yellowish opaque purulent fluid mixed with blood is seen. It is much softer than red hepatization and if lifted will likely fall to pieces. In the abscess stage the lung may be a mass of abscesses or single ones. I have seen cases in which one of the lungs was a mass of matter held in by the membrane covering the lung. If the hepatization is extensive the animal usually dies before it has time to break down into pus, but if only one lung is affected the animal may live long enough for it to do so. Gangrene is not a common result of pneumonia in ordinary cases, but I have often found the lungs in this condition in pigs that died from hog cholera or swine plague.

Treatment: In no disease is it more important to make a proper discrimination in the treatment. The measures which would be beneficial in one case might cause a fatal termination in another. Animals of a vigorous constitution want reducing medicine, such as a dose of epsom salts, one to two ounces, and follow this by giving two to three drops of the fluid extract of veratrum viride every two or three hours, if it does not cause vomiting. There is no medicine in the materia medica which I find so useful as veratrum viride in pneumonia in the pig if it be in a vigorous condition. To prevent vomiting give from six to ten drops of laudanum a short time before giving the veratrum. Very often if the case is seen early it will recover without any further treatment, but, should it prove obstinate, it will be proper to resort to calomel; two to three grains of calomel should be given with one grain of opium twice a day. It will be found useful to add half a grain of ipecacuanha to the above; should it not agree with the stomach leave it out or reduce the amount to one fourth of a grain. It is important to push the mercury until the gums become sore, then stop. In the declining stage of the disease expectorants should be given, and nothing will be found more useful for this purpose than the compound syrup of squill in doses of from twenty to thirty drops two or three times a day. If the breath should become fetid the spirits of turpentine in twenty to thirty drops three or four times a day should be used. Should the animal become weak give stimulants and tonics; carbonate of ammonia in five to ten grains made into a pill three or four times a day will be beneficial, two tablespoonfuls of whisky and two to four grains of quinine, also cod liver oil, brandy and eggs. Feed on anything the animal will take, such as new milk and a little otmeal gruel and all the cold water it wants. When gangrene sets in, which can be easily known by the fetor of the breath and a discharge of a dirty, foul-smelling matter from the nose, treatment is useless in the pig.

PLEURISY (INFLAMMATION OF THE PLEURA.)

This is an inflammation of the pleura which lines the cavity of the chest. This is one of the most frequent of the inflammatory diseases and the pig comes in for its share of it, as it is a common occurrence to find, on removing the lungs of a pig that had been in good condition and had been killed, a part of the membrane adhering to the walls of the chest, caused at some period more or less remote by an attack of pleurisy more or less severe. A pig may be affected to a slight extent and no notice be taken of it. It is only when the animal has a severe attack that our attention is drawn to it by the pig being off its food and appearing sick.

Causes of pleurisy: Cold, wet weather in the fall and spring and the pigs having wet beds to sleep in, cold east winds and the pigs not having a comfortable place to rest and get warmed. It is also a result of some other disease such as rheumatism and also from an injury such as a kick from a horse, etc. Some seasons it is a very common disease and hundreds of pigs die from it and its true nature has not been known by the owner of the animal.

Symptoms: The disease usually commences with a chill and a sharp pain in the side, which often causes the animal to walk lame on the foreleg of the affected side, or if both sides are affected the animal will move stiffly. There is a short spasmodic cough causing much distress. The breathing is fast and short and the abdominal muscles are brought into play to help to expel the air, as the pig keeps the walls of its chest as quiet as possible while suffering from this disease. The pulse is fast, small and hard and the temperature will be elevated, indicating fever. If the ribs are pressed upon the animal will evince pain and may squeal. It is not easy to judge the amount of pain a pig may suffer by pressure, as they will sometimes squeal on pressure when there is nothing the matter with them; still when other symptoms are present this one will assist. When there is much fever the pig will refuse its food, but may drink water. In bad cases it is very restless; it moves about with its back arched and stiff, and when it lies down it will be on its sternum. If the pig can be kept quiet and the ear applied to the chest in the early stages of the disease a crackling or crepitating sound will be heard at each movement of the lungs. If the acute stage is not relieved it goes on to the second or subacute; usually at this stage of the disease there is an outpouring of fluid into the cavity of the chest; as this increases the pain diminishes, but the breathing becomes more labored and difficult (dyspnea) and as the cavity fills up diminishing the breathing space the animal soon dies from suffocation. When effusion takes place the pain is reduced and the animal may appear better for a day or two, but soon a low form of fever sets in and if the animal is not relieved will die. On applying the ear to the affected side or sides the friction sound first heard will have disappeared and we may be able to detect a light splashing sound and a partial or complete loss of respiratory murmur, which may enable us to detect how high the effusion has risen. Very frequently there will be flakes of lymph, and sometimes false membrane will form. I have seen some cases that have died when one could collect handfuls of this false product.

Termination: Simple pleurisy of one side usually terminates favorably if treated properly and in the early stages. When effusion has taken place the chances of cure are less, but may generally be effected in cases where it is not complicated with other diseases. When it occurs in the course of febrile affections or in pigs debilitated from some other cause it almost always increases the danger and often hastens the fatal issue. Pleurisy is often present in cases of hog cholera and no doubt it helps on the fatal termination of that disease.

Treatment: First—Give from one to two ounces sulphate of magnesia in half a pint of cold water and follow this with fluid extract of veratrum viride one to two drops mixed with fifteen to twenty drops of tincture of hyoscyamus every four hours until the pulse becomes softer; after the physic has operated if the cough is troublesome and the animal seems to be in pain give one grain each of powdered opium and ipecacuanha in a pill every four hours; also give from ten to fifteen grains of nitrate of potassium three or four times a day to cool the system and keep the kidneys active. If the inflammation should continue after the second day combine two grains of calomel with the opium and ipecacuanha. After the decline of the fever should effusion still remain give syrup of squills with digitalis, a dram of the former and two drops of the fluid extract of the latter at a dose three times a day and apply a cantharidal blister to the sides. If the system is in a depleted condition it will be necessary to give tonics and stimulants; nothing is better for this purpose than tincture chloride of iron in dram doses with two to four grains of quinine three times a day in a little syrup; if the cough should continue give the opium as above. I have found benefit from the use of the compound syrup of squill in thirty drop doses three times a day. Also give iodide of potassium in ten grain doses between times. The animal's strength should be supported by milk, eggs and whisky; the whisky acts as a stimulant and aids digestion.

EMPHYSEMA OF THE LUNGS.

This name has been applied to that affection of the lungs in which their tissue is morbidly distended with air. There are two varieties of this derangement, one in which the air cells are distended, and the other in which the air has escaped from the cells into extravesicular or interlobular areolar tissue or upon the surface of the lung beneath the pleura. This complaint is not uncommon in the pig; but is not often noticed until after death.

Causes: Emphysema is often the result of over exertion as when a pig is pursued by a dog or man the increased amount of air taken in overfills the air cells, which distends them to such a degree that they may rupture and thus allow the air to extravasate into the lung tissue, or the air cells do not contract to their natural size. It is also the result of lung disease.

Symptoms: Slight emphysema cannot be detected in the pig until after death, but when so considerable as to produce observable effects it is always attended with dyspnea (difficult breathing), which is very distressing to the pig and is often associated with thumps, or rather it is often mistaken for thumps. Pigs often die suddenly from emphysema. Cough is sometimes present; during severe paroxysms there may be an expectoration of a clear fluid or it may be frothy; the animal will stand during the intervals of the paroxysms with its mouth open and turning its head to one side with the nose pointed upwards. In post mortems that have been made of animals that died of emphysema we find that the lungs do not collapse on opening the chest, but sometimes on the contrary expand, as if previously compressed by the ribs and diaphragm. This is the result of the inelastic or rigid condition of the membrane which forms the air cells. In consequence of this rigidity they do not contract upon the air which they contain and therefore remain distended. The lungs are very light and do not sink in water as in the sound state. It crepitates less upon pressure, has a firm feeling, and pits under the finger. The surface of the lung will be found uneven on account of some parts of it being more distended than others. Some of these projections may be single and not larger than a pea; that these are distended air cells is proved by the circumstances that they cannot be moved from place to place under the pleura by pressure. When a diseased lung is cut into the air cells are found to be in various degrees enlarged generally to about the size of a millet seed, sometimes to that of a hazel nut and sometimes larger. The small ones are merely dilated vesicles, the larger are produced by the rupture of the intervening coats and the gradual absorption of the torn walls of the cells. The dilatation may affect only one or a few cells or may occupy isolated spots as single lobules; for example, while others remain unchanged or may extend to large and continuous portions of the parenchyma it may be confined to one lung or both may be affected. When only one lung is affected it will be found to be so much larger than the other that it will displace the mediastinum and heart. The dilatation most frequently affects the anterior and its borders than elsewhere. The edges of the tubes are sometimes fringed with the projecting dilated vesicles of different sizes. The small bronchial tubes are usually also dilated in the emphysematous parts. In consequence of a diminished supply of blood to the affected parts of the lung it will have a whitish appearance which will contrast with the healthy portion. The lung is also less moist than in health.

Treatment: Keep such an animal as quiet as possible, and in the majority of cases it will take on fat. If the paroxysms of dyspnea are severe give from one to two drams of Hoffman's anodyne in a little cold water every half hour until relieved, or twenty to thirty drops of tincture of opium in a little water every hour until relieved.

CHRONIC COUGH.

Pigs are very subject to cough, which in a great many cases does not seem to affect the health of the animal.

Causes: Cough in pigs may arise from several causes, worms, indigestion, disordered liver and irritation of the membrane of the throat. If a number of pigs are affected with a cough at the same time it will only be a symptom of some other disease. If possible find the cause and remove it. If this cannot be done give the pig a dose of epsom salts and keep it in a warm, comfortable place, and if this does not relieve it give a teaspoonful of syrup of squill at a dose three times a day. If worms are suspected give one half oz. fluid extract of spigelia and senna at a dose every four hours until it purges. In cases of chronic cough caused by indigestion give a dose of salts and after it has operated give ten grains of sulphate of iron, ten grains of gentian and five grains of nux vomica at a dose in its food twice a day for two weeks. Another useful remedy is Fowler's solution of arsenic given in five drop doses three times a day in the food and continued for five or six weeks. Tar given in teaspoonful doses in the food three times a day is also good and as it cannot do any harm it is worth a trial.

BLEEDING FROM THE LUNG (HEMOPTYSIS.)

This is not a common disease in the pig and the only cases that have come under my observation have been caused by over exertion and usually have resulted in death; therefore, I look upon it as a very dangerous symptom in this animal. I have no doubt but that cases may occur in pigs from causes similar to those of the human family.

Causes: Over exertion from being driven too fast, fighting with each other, being kicked by horses, violent squealing and severe coughing, falls, etc. It may also be caused by disease of the air passages or the lung substance, bronchitis or pneumonia; in such cases the blood vessels have been weakened, and the extra quantity of blood in the parts causes the blood to extravasate into the lung tissue and air cells.

Symptoms: The blood may issue both from the nostrils and mouth of the pig, but most commonly from the nostrils. It is usually liquid, florid and more or less frothy, owing to the admixture of air. If the quantity issuing is great it will be less frothy. There will be more or less cough and if severe there will be a feeling of suffocation. If bronchitis should be present there may be more or less mucus mixed with the blood.

Treatment: The pig should be kept as quiet as possible. It often does more harm than good to confine a pig to give it medicine in such cases, and a slight hemorrhage is often beneficial to the animal in relieving congestion. If the hemorrhage should be severe give hypodermic injections of the fluid extract of ergot of rye in two dram doses; repeat every hour until stopped. Other remedies which are used with success in man cannot be of much service in the pig.

CHAPTER X.

DISEASES OF THE HEART.

Disease of the heart is of very rare occurrence as an independent disease in the pig, as their life is short and the great majority of them are sent to market before it has time to develop; but the heart is very frequently affected in conjunction with other diseases. Although I have had no well marked cases of this derangement in the pig and there has been nothing written on this subject, yet I have no doubt but that some of the sudden deaths which sometimes occur in the hog are due to some form of heart disease. The animal is usually in such apparent good health up to the time of death that no warning is given and it is only after making a post mortem examination that the true nature of the disease is discovered.

PERICARDITIS
(INFLAMMATION OF THE PERICARDIUM.)

Pericarditis is an inflammation of the capsule surrounding the heart and reflected over it. Endocarditis is an inflammation of the membrane lining the cavities of the heart. Carditis is an inflammation of the substance of the heart itself. As it is impossible to diagnose the one from the other in the pig I shall describe only one. I have met, in several cases in the pig while examining for other diseases, undoubted signs that the animal had had an attack of pericarditis at some former time, and there is no doubt but that at times when a pig shows signs of being a little off it may sometimes be this disease. It is often associated with such diseases as rheumatism and pleurisy.

Causes of Pericarditis: The most common causes are exposure to cold, direct violence, muscular exertion, such as running when a pig is fat and heavy, rheumatism, pleurisy and pneumonia and an impure condition of the blood, also hog cholera.

Symptoms: The attack is usually ushered in with a chill, which is often repeated (that is, in severe cases), and always followed by fever. The pulse is the most important symptom. In the early stages of the disease it is usually full and somewhat irregular and as the disease advances it is exceedingly so, and on this account it can be distinguished from other inflammatory diseases. The different conditions of the pulse at different stages of the disease may be considered as indicative of the condition of the cardiac muscles, which are at first irritated into excessive action as shown by the strong, full, sharp, irregular pulse, subsequently weakened but still irregular, and lastly exhausted by excitement, so that the pulse becomes more and more feeble till the close. There will be more or less difficulty in the breathing, loss of appetite, although the animal may be thirsty. In some cases there will be severe pain indicated by the animal being restless or in some cases squealing. Little can be obtained from physical signs in the pig and we have to content ourselves with the other symptoms; but no great mistake can be made if we treat the inflammation and fever on general principles. Post mortem examination reveals effusion in the cavity; this usually begins about the third day and if the disease is not checked it increases until the end. There is also a fibrinous matter or lymph which is at first generally in the form of a soft, delicate film over the surface of the membrane from which it is easily separated; by degrees this becomes thicker and thicker over the surface of the membrane until in some cases it will be found to be an inch in thickness and somewhat hard. The occurrence of such changes must add greatly to the danger of the disease and indicates the need of prompt application of efficient treatment before it reaches this point. The changes which take place are not always the same. In some violent cases the disease has been known to run its course in a very short time and to terminate in forty-eight hours. In other cases the inflammation terminates in a few days before the effusion takes place. Sometimes mild cases may last for several weeks and the animal recover. When it is likely to prove fatal the animal becomes very weak and staggers about, refuses food, the pulse is small and fast—from one hundred and twenty to one hundred and fifty. There is usually some anasarca between the fore legs; an animal in this condition is liable to die at any moment from loss of heart power.

Treatment: In the early stage of the disease give from ten to fifteen grains of calomel at a dose, followed in three or four hours with from one to two ounces of castor oil. This will often cut the disease short. After the bowels have been freely acted on and the fever continues with a hard, strong pulse, give two to three drops of the fluid extract of veratrum viride every four hours until the pulse becomes softer and less frequent. Ten grains of nitrate of potassium given in a little water three times a day will be of service in keeping the kidneys active. If the pig should be in much pain give one grain each of opium and ipecacuanha three or four times a day in pill. If the bowels should become confined give one half to one ounce of sulphate of magnesia. After the fourth day the animal should get ten grains of iodide of potassium three times a day and if the heart is very irregular give two to three drops of the fluid extract of digitalis three times daily. If the animal becomes weak its strength should be supported by stimulants, such as carbonate of ammonia in from five to ten grain doses three times a day or whisky and milk. During the disease give the pig all the sweet milk it will take, with some roots or fresh clover, if it can be obtained. A little oatmeal gruel will also be beneficial. Keep the pig in a comfortable place.

HYPERTROPHY AND DILATION OF THE HEART.

Hypertrophy and dilatation of the heart are not readily detected in the pig while it is alive. I have never detected a case of it. A pig may have this trouble and still take on fat, and as it is a disease which is slow in producing bad symptoms, the pig is sent to market before it would likely be detected. I have seen some hearts of pigs which were killed for pork and also died from other diseases which had enlarged or dilated hearts. If we did detect such a case treatment would be of no service.

FUNCTIONAL OR NERVOUS DISEASES OF THE HEART.

The only one in the pig which will come under this head is palpitation of the heart (thumps). This term is given when the pulsations of the heart are inordinate and can be perceived by the observer. It is a very common and fatal disease in the pig.

Causes: Palpitation is rather the result of a disease than a disease itself. It is usually caused by an unhealthy state of the blood which may be either too rich and abundant, causing plethora, or have too much water or be otherwise depraved, as in anemia. The true causes are therefore such as produce either of those conditions. It may also in some cases be caused by a deranged state of the stomach. It is frequently the result of rheumatic irritation affecting the heart through the nerves. Worms in the intestines will sometimes produce it. The most frequent cause of this disease in the pig is a morbid condition of the system, the result of feeding too much corn or food deficient in nitrogen.

Symptoms: The pulsations are increased in frequency or in force or both; they are sometimes regular, but more frequently irregular; the pulse partakes of the same qualities. The pig is seen to stand and its body is jerked forwards and backwards every second or so. In some cases it is very slight and in others very violent; the sound is loud and can be heard at some distance. It sounds as if some one was striking the pig on the inside, causing the whole body to shake. I have caught the hog so affected to try and find out whether it was the heart or the diaphragm that was affected, but the animal struggled so that an examination could not be satisfactorily made; but I have no doubt but that in a number of cases both the heart and diaphragm are implicated in the affection. It is not of much importance, as the treatment would be the same in both cases.

Effects: If this disease is not soon checked injurious effects are sure to result from the irregular supply of blood consequent on this affection. After death there is usually found considerable effusion of frothy blood around the heart and between the lungs and diaphragm, and the other tissues and organs of the body are changed in appearance; they are soft and flabby, with more or less congestion, especially is this the case on the surface of the lungs.

Treatment: I have found a very successful treatment for this disease. In the early stages give one to two ounces of epsom salts or one to two ounces of castor oil; follow this by giving from fifteen to twenty drops each of tincture of opium and digitalis every two hours until the animal is relieved, and in the great majority of cases this can be done in from twelve to eighteen hours. In severe cases the dose can be increased or given oftener. Hoffman's anodyne in dram doses given in a little cold water every two hours in cases where the animal is much exhausted will be found very useful. If this cannot be obtained give two drams aromatic spirits of ammonia every hour in a little cold water. As soon as the pig will take food feed it on new milk and a little oatmeal gruel until it regains its usual health.

CHAPTER XI.

DISEASES OF THE ARTERIES AND VEINS.

These are of very rare occurrence in the pig and as they usually take a long time to injure the health of this animal it can be fattened and sent to the market before it causes any trouble; therefore, I will not take up the reader's time describing them.

CHAPTER XII.

APOPLEXY.

This derangement is almost entirely confined to pigs that are too fat and are getting too much to eat.

Symptoms: If the animal is eating it will stop suddenly, appear restless, stupid in its movements; the eyes become prominent and blood-shot; it foams at the mouth, falls over and may die in a few minutes; others may fall over apparently dead, and in a few moments revive and get well.

Treatment: Bleeding from the veins of the legs is recommended. Tie a string tight above the knee of the forelegs; the vein on the inside of the leg will be seen to fill up; then take a sharp knife and make an opening sufficient to allow a free stream of blood; serve the other one in the same way. After removing one and a half pints of blood, pin up the wound in the skin. Press a small pin through the skin at the edges of the wound and tie a thread around the ends of the pin, thus closing the wound. Let the pins remain in until the skin is healed. Hogs that are over fat and show any signs of dullness should get a dose of epsom salts —two to three ounces—and be fed sparingly for a few days, until the dulness wears off.

PLETHORA.

This is a morbid increase of blood beyond the wants of the system. It is not, however, a mere augmentation of volume in the circulating fluid that is entitled to this name. This may result from an excess of the watery ingredient and is not incompatible with the state of the animal which is opposite to that of plethora. There must be a morbid increase of those constituents of the blood upon which its nutritive and stimulating properties depend and to which it owes its peculiar character, such as the red corpuscles, fibrin and albumen. There may or may not be any increase of bulk. It is not necessary that there should be an absolute increase of the blood in order to the existence of the disease. The quantity may remain precisely the same and yet if the wants of the system for the support of its various functions should diminish the phenomena of plethora may result; for it is the loss of balance between the supply and consumption, the former being in excess, that constitutes the affection. (Wood).

Causes of Plethora: A loss of equilibrium between the supplying and expending processes is the immediate cause of plethora. Digestion and absorption are relatively more vigorous than nutrition and secretion. Thus feeding pigs more than they ought to be fed will produce this effect, especially if the animal has little exercise. In some cases digestion and absorption may be healthy while nutrition and secretion may be in a deranged state. Some animals have a peculiar tendency to the overproduction of blood and become plethoric without any assignable cause.

Symptoms: Pigs affected with this disease are stupid, lie around, and if made to rise will show signs of vertigo, stagger and in some cases fall. There is often palpitation of the heart and oppressed breathing; the lips, tongue and conjunctiva are red and sometimes somewhat swollen; bleeding from the nose sometimes takes place. I have known some few cases in which the pig was very fat and at the same time affected with this disease in which the animal lost the use of its legs. The pulse is full and somewht accelerated.

Treatment: Regulate the diet. Give milk and oatmeal in small quantities and turn the pig out so that it will get an opportunity to move about, and in some cases it should be compelled to move around for a short time each day in order to stimulate secretion. Give one to two ounces of epsom salts and after this operates give bicarbonate of potassium in two dram doses three times a day. If the animal loses the use of its legs they should be rubbed several times a day.

ANEMIA.

Anemia is a morbid deficiency or poverty of the blood, caused by lessening of the nutritive constituents of the blood and an increase of the watery parts. Young pigs get into this state without any apparent cause.

Symptoms: There will be puffy swellings around the eyelids, between the wings of the lower jaw and between the fore legs. The animal becomes unthrifty, although it may eat well. The mucous membrane of the mouth and eyelids becomes pale. The pig is very often troubled with a cough and a discharge from the nose. I have seen young pigs affected by this derangement have fits and when the cause of the disease was removed and the disease itself properly treated the fits disappeared. There is no doubt but that the nervous system suffers more or less from anemia.

Treatment: Remove the cause if possible; this is often accomplished by giving a complete change of diet. It will be advisable to give one to two ounces of epsom salts or one to two ounces of castor oil to clear out the stomach and intestines, then give from ten drops to one dram of the tincture chloride of iron in the food three times a day. A dessert spoonful of cod liver oil three times a day is very useful. Allow the pigs to run out. This is usually all the treatment that is necessary.

CHAPTER XIII.

PURULENT INFECTIONS OF THE BLOOD.

By this is meant a diseased state of the blood attended with a low form of fever dependent on the absorption of poisonous material from the decomposition of tissue into pus. This sometimes takes place after castration or other wounds that the animal may receive. There is a tendency in this disease for abscesses to form in various parts of the body; they are especially liable to form on the lungs and liver. In pigs which have died of this disease I have found a number of abscesses, varying from the size of a pea to that of a walnut; these abscesses may be found in any part of the body, even in the muscles. These purulent collections are usually surrounded by more or less inflammation. There is also a tendency to a low form of erysipelatous inflammation in various parts of the body, ending frequently in gangrenous abscesses, which usually terminate fatally. The symptoms of this disease are not well marked at first in the pig. The animal refuses food and has shivering fits; it lies around and does not want to get up; if it does so it will act stupidly, breathe fast, the pulse will be fast and weak, the tempera-

DISEASES OF THE HOG. 135

ure will be from one hundred and five to one hundred and six; this low form of fever may be all that is noticeable, but if these symptoms should take place after an operation or an injury there need be no mistake. In some cases there will appear abscesses of the external part of the body having a tendency to gangrene, or they may heal, but others will make their appearance and so on until the animal dies.

Treatment: This should be of a stimulant and tonic nature, such as good whisky or brandy or carbonate of ammonia in from five to ten grain doses three or four times a day. The tincture chloride of iron in dram doses well diluted three times a day is one of the best. Nitro-hydrochloric acid in doses of from ten to fifteen drops well diluted three times a day is also good. The strength should be kept up by giving new milk and eggs beaten up and mixed with a little whisky and the pig kept in a dry, comfortable place. If abscesses should form they should be opened as soon as ripe and dressed with peroxide of hydrogen one part, water two parts. If the sores become gangrenous apply a little terchloride of antimony to destroy the unhealthy parts; then use oxide of zinc one ounce, vaseline two ounces; use a little twice a day.

CHAPTER XIV.

ANTHRAX OF THE PIG.

Anthrax. There are a number of diseases of the pig which come under this head, and all are of a carbuncular nature; that is, it is a constitutional fever at first affecting the finer tissues of the body and finally locating on some particular part, causing either large swellings or pustules on the part affected, and if there is a tendency to the disease prevailing swine are sure to come in for their share of it. In this affection the blood is darker than in health, and the name "charbon" has been given to it by some writers. This condition is caused by a bacillus called the "Bacillus Anthracis," a very large bacterium. This vegetable organism gains entry into the blood, and there multiplies by rapid reproduction. It is imagined that it wars with the red corpuscles for the possession of oxygen, and soon all the higher tissues become implicated. It is said that heat and moisture are favorable for their development and that the disease is more common in the spring and fall when the weather is warm and moist. Pasteur made a number of experiments and came to the conclusion that the bacilli gained entrance into the system with the food, and wounds about the mouth were the points of entry. A number of experiments I made with these bacilli were by no means satisfactory, unless injected directly into the tissue. I have seen outbreaks of this disease in cold weather when everything was in an unfavorable condition for the development of the bacilli. I consider, from what has been done, that it has been proven that the bacilli will cause the disease, but how it gains entrance into the animal body has not been satisfactorily proven. It will make its appearance on a farm where everything is favorable for the development of the bacilli, and again at other times it will appear in a very virulent form where everything is kept in the best hygienic condition. The study of this disease and the manner in which the outbreaks take place are very conflicting. There is room for much more work in this line. There are four forms of this disease in the pig. First, aphthous fever; this form of anthrax commences with loss of appetite, uneasiness, trembling, anxious and staring look, hot mouth and an increased flow of saliva. Early in the attack pustules appear on the inside of the lip and margins of the snout; they are not numerous, but there is considerable inflammation surrounding them which causes the parts to swell. The vesicles are at first white, then change to a brownish color or in some cases black. They generally extend up the nose, which is somewhat swollen, giving the animal an ugly appearance. Shortly the vesicles burst and the tissue beneath sloughs, leaving more or less of a cavity resembling an ulcer. At other times there may be a slight elevation of a fungous nature. I have seen a few cases in which the tissue sloughed clean from the bone. The breath becomes fetid, usually a foul-smelling diarrhea sets in, mixed with blood, followed by great prostration, the animal dies in from twenty-four to forty-eight hours from the beginning of the attack; some few cases may last longer. The disease is contagious and all the well pigs should be removed at once from the affected ones and from the locality. All those which die should be burned or buried deep, with a good coating of fresh lime over them.

Treatment: In the early stage give one to two ounces of epsom salts at a dose, this to be followed by ten to fifteen drops of the oil of gaultheria in a little syrup or sweet oil or a solution of gum arabic three or four times a day. If the animal is weak, do not give the salts, but give the oil of gaultheria and one dram of tincture chloride of iron three or four times daily. If there is foul-smelling diarrhea I have found great benefit by giving ten grain doses of boracic acid, three or six times a day; I have given as much as two drams in divided doses in twenty-four hours; it is best given in a tablespoonful of sweet oil or cotton seed oil; if the animal seems much distressed one to two grains of powdered opium can be added to the acid. The mouth and affected parts should be washed several times a day with peroxide of hydrogen (fifteen volumes), one part, water two parts. When the pustule sloughs apply a little terchloride of antimony to it to destroy the diseased tissue; it can be repeated every third day if necessary. I have had good success from this method of treatment. The pig should be fed on milk and if it will not take it it should be fed with a little milk and beaten eggs, with a spoonful of good whisky in it to help digestion.

NECK ANTHRAX.

This form of anthrax is characterized by an eruption of boils which appear on the back of the neck over the parotid glands below the ear. The bristles on the affected parts stand erect; they are dry and stiff, and if touched or pulled the pig will squeal with pain; the skin is discolored, usually of a purple tint; there is high fever, loss of appetite, thirst, grinding the teeth, and a hot clammy mouth; there is difficulty in swallowing and oppressed breathing, which would indicate sore throat; the affected parts usually slough, erysipelas sets in and the animal generally dies about the ninth or tenth day.

Treatment: Give one to two ounces epsom salts to cool the system; then give five to eight drops of the strong nitro-muriatic acid at a dose diluted in a wineglassful of water three or four times a day, continue this for three days, then give two to four grains of quinine and two tablespoonfuls of whisky in a little water three times a day for two days, then give from twenty to thirty drops of tincture chloride of iron at a dose in water three times a day. Bathe the affected parts three times a day with a lotion made by mixing one half ounce each of acetate of lead, sulphate of zinc and carbolic acid in a quart of soft water. Feed the pig on oatmeal and milk. If this form of treatment is faithfully carried out a great many of the affected animals will recover.

GANGRENOUS ERYSIPELAS.

This is another of the anthrax diseases which affect swine; it has been called "wild fire" in the Old World. This derangement begins with a low form of fever, the animal appears dull, does not like to walk or stand, but lies buried in the straw or anything it can get into; the temperature is high, the pulse fast and weak, and the breathing quick and short; red spots appear on the breast, belly and inside of the legs, and often cause swelling of the skin, and later on it becomes dry and loose, as if it was much too large for the pig, and crackles on pressure. In some cases the greater part of the skin becomes affected; it will first be red, then become purple, and some parts in the later stages of the disease may be black in spots. In others it forms ridges and cracks. The hair drops out in patches and the animal soon becomes weak and cannot stand on its hind legs, and usually dies in a week or ten days from the beginning of the attack.

Treatment: In the early part of the disease give ten grains each of aloes and calomel at a dose in a little gruel. If this does not cause purging in ten or twelve hours repeat the dose. Follow this by giving twenty to forty drops of the syrup iodide of iron in a little syrup three times a day, also give four grains of quinine in a little whisky and water twice a day for two days. In the early stage of this disease the skin should be kept wet with a lotion composed of acetate of lead one ounce, water one quart; when the skin becomes dry and cracks rub it once a day with an ointment made by mixing two ounces oxide of zinc with four ounces of vaseline. Feed as directed in the former. The pig should get all the cold water it can drink in all cases of fever.

GANGRENOUS ANGINA.

This form is common in the pig, its principal seat being the throat and is characterized by a difficulty in swallowing and breathing, with a wheezing sound; the tongue often swells and becomes of a dark purple or of a bluish black color, "black tongue," there is usually a painful swelling on the outside of the throat, sometimes extending down between the fore legs. If the skin is white it will be changed to red and later on dark. As the disease progresses the breathing becomes more difficult, the tongue may swell to such a size as to cause suffocation.

Treatment: If the animal can swallow give it one to two ounces of epsom salts; follow this by giving five to eight drops of the strong nitromuriatic acid well diluted three or four times a day. If the tongue is much swollen take half an ounce of acetate of lead and one ounce tincture of opium and mix in a quart of water, shake up well, then take a piece of sponge and tie a cord to it, wet this sponge with a little of the lotion and press it well back on the tongue, let it remain for a minute then withdraw and repeat, and in this way we can sometimes succeed in relieving the swelling. I have saved the life of several pigs by continuing this plan of bathing for several hours. It is good practice to inject into the swelling on the outside a lotion composed of peroxide of hydrogen, one part to four of water. This is a powerful antiseptic and will sometimes stop the process of the disease. In treating the above diseases it should always be borne in mind that if any of the matter from the affected animal gets into the blood of man he runs a great risk of contracting the disease.

CHAPTER XV.

SKIN DISEASE.

The skin on the animal's body serves as a protection to the soft structures beneath it, also to prevent noxious materials from passing into the tissue beneath it; it also allows the escape of waste substances from the body. There is very little perspiration exudes through the pores of the hog's skin and therefore this animal suffers from heat and seeks the mud puddle or water hole on a hot day to cool himself. If a pig is driven fast on a hot day it will be seen to open its mouth and protrude its tongue in order to cool off. On this account pigs should be sheltered from the heat of the sun and if possible should have a good clean mud puddle to wallow in; what is meant by clean mud is where the water in it is fresh, not stagnant or of a yellowish green color; such muddy pools are full of microbes which may in some cases find their way into the animal's body and cause disease. Still, from practical experience, I think the heat of the sun will do more harm to the health of the hog than wallowing in a stagnant pool. If from disease of the skin or other causes the pores should become blocked up the animal is liable to suffer more or less from the effects of the effete matters being penned up in the system, and therefore valuable sows and boars should have their skins thoroughly cleaned occasionally, thus preventing disease and keeping off vermin.

CANKER OF THE NOSE AND FACE (SORE NOSE.)

This is a form of skin disease which usually affects the nose first and by degrees spreads up the face and sometimes over the ears and in patches on the body.

Causes: It is the result of a parasite which resembles that of mange, but seems to have poisonous properties. As it spreads the part becomes of a cankerous nature, causing constitutional symptoms and death. It is not as yet known where this parasite originates, as I have found it affecting pigs where the sanitary conditions were good. I have made a number of experiments to find the origin of the parasite, but have found nothing satisfactory.

Symptoms: The first thing noticed is a contraction of the skin which wrinkles, giving the nose a short, stubby appearance. The animal rubs its nose on the earth, snuffles and runs as if it wanted to get away from something. By degrees it breaks out in sores, which may extend up to the eyes and ears and in some cases form hard patches on the belly. I have seen cases of long standing where the parasite buried deep into the muscles of the face so that a slough would take place clean to the bone of the jaw, leaving cankerous edges to the sore. The conjunctiva of the eyes and root of the ears may be also implicated. Pigs affected with this disease do not thrive and often die.

Treatment: All the well ones should be removed and their heads rubbed with a liniment made by mixing one ounce of carbolic acid in half a pint of raw linseed or cotton seed oil. The affected ones in the early stages, before a slough takes place, should be well rubbed with an ointment made of iodine one-half ounce, vaseline eight ounces, or ichthyol one part, vaseline two parts. Repeat in three days. If this is well done it will stop the disease by killing the parasite. After it sloughs and becomes cankerous apply a little terchloride of antimony to the sores with a feather. Let it alone for three days and if the parts look healthy use a solution of tobacco one part to thirty parts of water, and apply a little of this; be sure that every part is wet with it, then apply a little common tar to the affected parts. Take notice of the pigs; if they seem easy let them alone, but if they should be restless wash the parts with warm water and soap and use the tobacco solution again, and so on until the animal is cured.

MANGE.

This is also a parasitic disease, but not so difficult to cure and seldom causes death. It is caused by a parasite which burrows under the scarf skin, producing considerable irritation, destroying it so that scabs form, and on account of the great itchiness on the part the animal rubs it, causing the part to become a rough sore. This disease is contagious from one pig to another. It generally appears first on the thin parts of the skin under the arm, behind the ear, inside of the thighs and upon the back. This disease in the early stage resembles eczema, but if the eyesight is good and assisted by the bright sunlight, the parasites may be seen as a moving white speck, but can be readily seen by a small magnifying glass. The cause of this disease is by contagion; that is, the parasites themselves or their eggs must get on the skin in order to produce the disease. It is therefore necessary that all the unaffected pigs should be removed from the premises where the diseased ones are or have been, and the wood work of the sties washed with a strong solution of carbolic acid.

Treatment: Wash the pigs all over with soft soap and water, then rub in well dry sulphur. When the sulphur comes in contact with secretions from the sores, it forms a compound poisonous to the parasites. In a day or two give the animal another washing, and rub on some more sulphur. A very sure remedy is to boil for one hour two ounces of stavesacre seeds in one and one-half quarts of water, and keep it nearly boiling for an hour longer; make up the water to the quantity originally used. Such a solution, rubbed into the skin, not only kills the parasite, but its eggs also. Repeat in a week, if necessary. Another is to steep one part of tobacco in twenty parts of boiling water for a few hours, and, after washing the pig, apply a little of this to the affected parts with a sponge.

If the surface to be covered is large, only apply it to one part today and to the other tomorrow. For instance, if the neck and the legs are affected, apply to the neck first, then to the other parts the day following, and there will be no bad results from the use of the tobacco, and it is an excellent remedy when carefully used.

LICE.

The causes of lice in pigs are bad food and filthy sties. When pigs are badly housed or kept, lice will likely appear and will spread from one pig to the other. The louse of the pig is of a dusky iron color on its back, and gray or ashy yellow on the belly, and has long legs. Lice are a great torment to the pig, and it will be impossible to fatten a hog which is lousy, and they will prevent young pigs from growing. There are a number of remedies for the destruction of lice. The best, if attainable, is to boil two ounces of stavesacre seeds in three pints of water, let it nearly boil for an hour longer, making up the water to the quantity originally used. Wet the animal all over with a little of this. It will kill the lice and the nits. Solution of tobacco one to twenty of water, is very effectual in destroying lice, and when used with caution there is no danger. Cotton seed or raw linseed oil two parts, kerosene one part, this to be rubbed over the animal. The only objection is, it being a greasy substance, it sometimes clogs up the pores, and on that account either of the first two are preferred.

WARTS.

These are caused by a deranged condition of the scarf skin, but subsequently the true skin becomes affected, thus producing on the surface of the body growths of various sizes. When these are rubbed or bruised they ulcerate, and thus form fungus-like masses, projecting from the skin. They are very unsightly, and bleed on the slightest pressure. In some cases they are flat, spreading over the skin.

Treatment: For those which project from the skin and have a neck, tie a piece of sharp cord tightly around the base. If it does not slough off in a week, tie on another. When it sloughs off touch the part with terchloride of antimony once a day for a few days, to destroy the roots of the wart. In cases where they are flat, scrape off the scurf until the blood begins to ooze, then apply a little of the antimony to it with a feather. On the second or third day remove the scab that forms, and apply a little more of the antimony, and so on until it is lower than the surrounding skin. Then apply zinc oxide one ounce, vaseline two ounces; use a little once a day to heal it. If it should show signs of growing up again, use the antimony as above.

URTICARIA, NETTLE RASH, SURFEIT.

This is a non-contagious disease characterized by the cutaneous elevations which are surrounded by redness, which can be seen if the pig's skin is white. It is usually the result of a deranged digestion or of feeding too long on one kind of food, or too much of any kind of starchy food. Some pigs are peculiarly susceptible to this disease.

Symptoms: The sudden appearance of elastic prominences on the skin, accompanied by great itching of the parts; it may pass off as suddenly as it appears. There is a second form of this disease in which the lumps may rise on any part of the body and if they do not pass off soon, may form vesicles and discharge a glutinous fluid, and the animal may appear dull, the appetite may be somewhat impaired and the animal's health disturbed.

Treatment: Give one to two ounces epsom salts and change the food. To relieve the itching mix one ounce of acetate of lead, one ounce tincture of opium in a quart of water and bathe the parts well with a little of it three or four times a day, if necessary. If the animal is not cured in a few days give five drops Fowler's solution of arsenic in its food three times a day. The trouble usually disappears after the stomach and bowels have been cleaned out.

LICHEN.

This is a form of skin disease consisting of pimples about the size of millet seeds; they develop principally around the hair follicles in patches; the hair falls off and the skin remains bare for five or six weeks, when a layer of scales drop off, and then the hair begins to grow. This malady is apt to recur.

Treatment: Give the pig a teaspoonful of sulphur and ten grains of nitrate of potassium at a dose three times a day in the food and continue it for two weeks, if necessary. The skin should be well washed with a solution made by dissolving one ounce of carbonate of potash in a quart of soft water, then use clean water to wash it off; repeat the washing once a week.

PRURIGO.

This is an eruption of pimples having nearly the same color as the skin and attended with excessive itching. The eruption is often confined to one spot and sometimes it attacks several parts of the body at the same time. The parts most usually affected are the neck and shoulders.

The only symptom of this disease is the constant excessive itching. From the want of color the pimples are not observed, but the animal rubs itself so severely that the skin often becomes abraded, sometimes small black scabs may be seen. By running the fingers over the affected part the skin will feel rough or the pimples may be detected. It may occur at any period of the animal's existence.

Treatment: As the majority of skin troubles are the result of faulty digestion the condition of the stomach should be attended to. Give from one to two ounces epsom salts and change the diet. If the animal is in a debilitated state give thirty drops syrup iodide of iron at a dose in a little water three times a day after meals, or five drop doses three times a day of Donovan's solution of arsenic. Bathe

the affected parts three or four times a day with a lotion made by mixing half a dram of hydrocyanic acid to the ounce of water. Twenty drops of creasote to the ounce of lard rubbed on the parts once a day is very useful. In chronic cases use a solution of corrosive sublimate, three grains to the ounce of water. If the skin is hard I have found an ointment made by mixing one ounce of ichthyol and two ounces of vaseline to be very useful. The ointments of oxide of zinc, one to two, or iodine one to eight of vaseline are also beneficial.

PEMPHIGUS.

This disease of the skin is characterized by bladders or elevations of the scarf skin varying from the size of a pea to a walnut, containing a yellowish transparent fluid and terminating in the formation of a scab. This disease usually occurs without fever. The parts usually affected are the neck, sides, back and sometimes the outsides of the legs. They usually remain for three or four days, then break, form a scab and heal. It occurs most frequently in young pigs, but has appeared in adult hogs. The causes of this disease are obscure. It is supposed to be the result of an impaired condition of the system and exposure to the hot sun.

Treatment: Give one to two ounces of epsom salts and follow this by ten grains of nitrate of potass three times a day in the food. A complete change of diet is sometimes all that is necessary. After the blisters break and discharge their contents apply a little oxide of zinc ointment or a simple cerate to protect the sore from the air.

RUPIA.

This is a disease which resembles pemphigus, but the elevations are flatter and contain a dark colored fluid; they are followed by a thick scab, easily separated and soon removed, or sometimes by ulcers. It is usually confounded with pemphigus, but as the treatment is about the same a mistake will be of no importance.

ACNE.

This is a pustular disease sometimes seen on the udder of the sow and inside of the thighs. It consists of small pimples which form on the skin; some of the larger ones may contain a little matter. In a week or ten days they begin to dry up, leaving a brown scab or mark. In some cases they become hard and of a reddish color and may remain in that condition for months. They make their appearance without causing any fever or itching. It does not seem to be contagious, as I have known a case in which the sow was nursing her pigs and none of them became affected with it.

Treatment: Give a mild dose of salts; follow this with syrup iodide of iron or five drops of Donovan's solution of arsenic three times a day in the food. Rub the affected parts with glycerine one ounce, tannic acid one dram, water one ounce. In chronic cases rub on a little soap liniment twice a day. If the pustules should suppurate and become sores apply the oxide of zinc ointment, made by mixing one ounce of the oxide of zinc with two ounces of vaseline. This can be used once or twice a day, according to the severity of the case.

SCALY DISEASES OF THE SKIN.

There are a number which are usually classed together as it would be impossible to distinguish one from the other in the pig, namely psoriasis, lepra, pityriasis, etc.

Causes: Heat of the sun in summer and cold in winter. This condition of temperature has a wonderful effect on the skin of some hogs. There are other causes which are obscure. The disease is not contagious.

Symptoms: The disease first begins by the formation of minute pimples too small to be seen on the skin of the pig; these pimples dry up and the scarf skin peels off in flakes; this is usually repeated several times and may pass away. At others the skin is inflamed, thickened, and intersected in all directions with furrows which are often deep and filled with a white powdery matter; the hair comes off and will not grow until the skin becomes healthy. Pigs affected with this scaly disease present a very ugly appearance, and it is the most common form of skin disease among hogs. In some few cases the skin will crack, causing much pain to the animal when it moves, especially if it be in the region of the joints.

Treatment: If the pig is fat give it from one to two ounces of sulphate of magnesia dissolved in half a pint of cold water; after this operates give acetate of potass in doses of half a dram three times a day in the food. The diet under the same circumstances should be of a cooling nature. Avoid corn and give fine ground oats and milk instead and some kind of green food. In weak animals give the same kind of food, but do not give a physic; instead, give from half to one teaspoonful of the tincture chloride of iron in the food at a dose three times a day; a tablespoonful of cod liver oil at a dose in the food three times a day will be found to be of great service in such cases. A dessert spoonful of the compound syrup of sarsaparilla at a dose three times a day is also useful. Keep the pigs out of the sun and wash the body well with soap and water and then bathe with a lotion of acetate of lead half an ounce, sulphate of zinc half an ounce, water one quart. In cases where the skin has become hard use glycerine two ounces, water two ounces, tannic acid two drams; apply once a day. Iodide of sulphur fifteen grains, lard one ounce, is also useful. When the skin requires a stimulant rub once a day with a little compound soap liniment. If the skin should crack the oxide of zinc ointment will be the best to use.

RING WORM (TINEA TONSURANS.)

This is not a common disease among pigs, although I have been called upon to treat a few cases. It depends upon the presence of a vegetable parasite which finds its way to the skin by contagion and develops rapidly when it finds a suitable place;

it may affect any part of the body, but most frequently the face and ears.

Symptoms: There appears a gray crust on the skin and the hair drops out; this keeps spreading in the form of a ring until the whole side of the face or ears are covered with it. The center parts become dry and hard, but the edges of the ring, if examined closely, are found to be very small vesicles, more or less moist. This disease is contagious to man to a certain extent; that is, it will develop for a time and then die.

Treatment: First remove as much of the crust as possible by washing with soap and water, using a brush, then dry the surface, and mix two drams of iodine with two ounces of vaseline and rub a little of this well in; repeat in a week if necessary; or mix carbolic acid one ounce, alcohol two ounces; apply a little with a small brush or a feather; repeat in a week, if needed.

SORE FEET.

Pigs which are kept on hard, slippery floors suffer from inflammation of the sensitive parts of their feet. I have seen some very heavy hogs suffer from the same cause on hard, dry ground. There are also cases of sores breaking out at the back of the hoof and between the toes. This is called "foul in the feet;" but I have not seen any cases of the contagious foot disease in this country, and the one to be described is a local disease caused by some substance irritating the parts at the heel or between the toes. I have often seen it caused by fine cinders of coal being scattered in the yard. This material gets in between the toes, causing irritation. Vesicles may form, burst, and discharge a yellow fluid, or matter may form. Unhealthy sores are the result of the irritation not being removed. Cinders from coal should never be put in yards where hogs, cattle or sheep are kept, as I have had a number of cases in cattle and sheep, as well as the hog, nearly ruined from this cause.

Treatment: In cases where the feet are tender and no sores appear, the animal should be kept for several hours on a bed of wet sand, as it is not practicable to poultice the feet of the hog, and the wet sand will answer the purpose. If the animal is very lame a dose of epsom salts, followed by ten grains of nitrate of potassium two or three times a day in its food will cool the system and help to relieve the sore feet. In a few days the animal is cured. If sores appear between the toes or at the heels, clean the parts well with warm water and soap to remove all dirt. If there is any proud flesh, which can be known by its bluish appearance and spongy aspect, apply a little terchloride of antimony with a feather once. If it has not removed all the proud flesh apply again on the third day. To heal the sores use chloride of zinc one dram, water one pint; dress once or twice a day, according to the severity of the case. Keep the pigs in a clean, dry place until the feet are well.

CHAPTER XVI.

DISEASES OF THE EYEBALL.

The eye of the pig does not seem to be subject to many diseases, but I have no doubt but that cases occur which are not noticed; I shall therefore only mention those I have met with.

Conjunctivitis: Simple ophthalmia results from injuries, and especially from foreign matters entering the eye. Exposure to cold, heat and acrid vapors, it is also the result of other diseases. When injuries and foreign matters are the cause one eye only is usually affected.

Symptoms: There will be a profusion of tears trickling down the cheek or cheeks, there will be a thickening more or less of the eyelids and a redness of their lining membrane; this also may be thickened, which nearly closes the eye. An examination of the eye must be made, the lids being separated by the finger and thumb and each lid inverted in turn. If the eye has been injured, for a day or two there will usually be a white scum over the corner, which usually disappears when the inflammation is reduced.

Treatment: If there is any foreign substance in the eye have it removed. This is not easily done as the rapid movement of the haw over the eye prevents one from getting hold of the offending body. Two drops of a three per cent solution of cocaine dropped into the eye will relieve the irritation and the motion of the haw. For all forms of simple ophthalmia the eye should be bathed with cold water for ten or fifteen minutes three times a day and after each bathing apply a little of the following lotion with a feather: Acetate of lead ten grains, tincture of opium ten drops, distilled water two ounces. I find this is the best of all eye lotions for simple inflammations of the eye, no matter what the cause may have been. After the inflammation has been subdued, if there should be any scum over the cornea mix five grains of nitrate of silver in one ounce of distilled water and apply a little once a day with a clean feather; this will stimulate absorption and the scum will soon disappear. The pig should be kept in a moderately dark place until well. If the inflammation should be very severe, give from one to two ounces of epsom salts at a dose; this will cool the system and act as a revulsent.

AMAUROSIS. (PARALYSIS OF THE RETINA.)

Amaurosis is a permanent dilatation of the pupil of the eye. The eye appears glassy, and in looking into the posterior chamber it has a greenish appearance and is completely impervious to light.

Causes: It is usually caused by affection of the optic nerve from some lesion of the brain. It is also caused by derangement of the digestive organs. I have been called to cases in which a number of pigs became suddenly blind and generally the brain was also affected. I have made post mortem examinations of some that died and the only lesion found was in the stomachs, which were inflamed and full of undigested food, and in some I think worms were the cause. The greater number were cured by giving a physic, epsom salts one to two ounces, or one ounce castor oil with two drops of croton oil in it; this was followed by giving two drops of the fluid extract of nux vomica and two grains of santonin at a dose in a little syrup three times a day; this was continued for a week or two when necessary. When worms are suspected, give half ounce doses of the fluid extract of spigelia and senna every four hours until it purges, then give the nux vomica and santonin as above. I have found this quite common among shoats some seasons, and if seen early it can usually be cured. There are other diseases of the eye of the hog, but they are not easily treated, nor have I found it practicable to do so. Usually they do not affect the health of the animal, which will take on flesh, and can be sent to the market.

CHAPTER XVII.

SCARLET FEVER.

This is a contagious disease, characterized by inflammation of the fauces (back part of the mouth) and a scarlet rash appearing usually on the second day and ending about the sixth or seventh. This disease is often confounded with measles. Although there is a marked difference in the human being it is not so easily distinguished in the hog unless the animal is white. If it should be mistaken for measles or measles mistaken for it, the error would not be of much consequence, as it has to be treated according to the stage and severity of the fever; that is, to confine the fever as much as possible and keep up the strength of the patient.

Symptoms: In the early stages there is languor, and stiffness caused by the muscles of the back being affected; there is a fast pulse, from one hundred to one hundred and twenty; high temperature, one hundred to one hundred and six; dry, hot skin, furred tongue, loss of appetite, great thirst, and great muscular weakness; sometimes the animal will vomit and the nervous system may be disordered, causing restlessness, delirium, stupor, coma or convulsions. Very often from the beginning there is inflammation of the throat and back part of the mouth, which, on being examined, will be found red and swollen; the tongue will be coated with a yellowish white fur, and projecting red pimples will be seen upon its surface, and is red at the edges and tip. If the animal has white skin about the face, neck, belly or inside of the legs, a red rash will be seen about the second or third day. In some cases minute pimples form, which are itchy and make the animal very restless. The bowels are usually constipated, but in some few cases there may be diarrhoea. In some cases the throat symptoms are very slight, but usually they are severe and occur before the rash and are very distressing, swelling both inside and out, and may prevent the animal from swallowing and make the breathing very difficult. This disease is readily mistaken for quinsy in the pig. The disease usually reaches its height in from five to nine days, and then, in the majority of cases, begins to decline. The rash fades; the dry heat of the skin diminishes; the pulse becomes slower; the throat symptoms disappear, and the tongue loses its fur and becomes clean, and the temperature is reduced; but in some cases it takes on worse forms, and an animal may die before the eruption appears, from shock upon the nervous system; or at any time during the attack from brain trouble or from inflammation attacking some of the vital parts, such as the lungs, stomach, bowels; or the animal may die from suffocation. The patient may sink from debility. As this disease requires to be treated according to the condition of the animal, changing the medicine several times daily, it makes it a very difficult disease for the farmer to treat. Professor Wood, in his practice of medicine, says: "In the vast majority of cases scarlet fever would end favorably without treatment; hence, the reputation acquired by homeopathy in this disease." Therefore, if such a disease as this should break out, keep the animal in a good comfortable place, keep the bowels open by giving one ounce doses of epsom salts, or a seidlitz powder occasionally. Some recommend to give diluted acetic acid, ten drops in a little water, several times a day. If there should be diarrhoea, give one ounce of castor oil and from ten to twenty drops of tincture of opium in it; repeat in ten hours if necessary. It is dangerous to check diarrhoea too quickly in this disease, as it is often an effort of nature to rid the system of poisonous material. If the throat is troublesome give ten grains of chlorate of potassium and three to five drops of fluid extract of belladonna in three or four tablespoonfuls of cold water three or four times daily. If the fever is very high and in the early stage of the disease, from five to ten drops tincture of aconite in a spoonful of water, will in some cases keep it down; but this should not be carried too far, as it is a very reducing medicine, and so is the disease, and there may be danger of collapse. If the animal is very weak and the pulse small, give stimulants, such as two teaspoonfuls of sweet spirits of nitre in a little water, three or four times a day, or two teaspoonfuls of good whisky in a little milk several times daily. Carbonate of ammonia is also good, given in ten to twelve grain doses in a little cold water three times daily. In great debility quinine five grains, sulphuric acid two drops, water one ounce, twice a day is useful. Cloths wrung out of boiling water and wrapped around the swollen neck, and continued for twelve to twenty-four hours, often relieves the distress.

MEASLES (RUBEOLA.)

This is a very common disease in young pigs. It is very contagious and is characterized by more or less cough or sneezing, red, watery eyes and also a watery discharge from the nose; the pigs are not so bright as usual and lie down a good deal; in mild cases the appetite is not much altered. In severe cases the throat becomes more or less affected and swallowing may be difficult. There occurs on the fourth day a red rash on the skin, first in minute pimples formed into distinct spots very slightly elevated above the skin; this rash is not easily made out unless the skin of the pig is white. The disease is usually mistaken for catarrh or cold, but by a careful examination the rash can be made out, as there are nearly always some white patches on most hogs. If the disease assumes a severe form the appetite is usually impaired, the animal is thirsty, the eyes are red and the eyelids swollen; if the pig eats anything it is usually rejected by vomiting. In some few cases the animal will have convulsions. There will be fever, fast pulse, hurried breathing, and if the tongue is examined it will be coated with fur, especially along the center. About the eighth day the disease begins to decline, the pig brightens up, the swelling of the eyelids becomes reduced, the cough is less frequent or may have disappeared altogether, the red color of the skin will diminish, the appetite return, and the pig will be quite well about the eleventh or twelfth day. Occasionally the pectoral symptoms at this stage increase instead of diminishing as they usually do, indicating that either bronchitis or pneumonia is setting in, and if it does so it will likely prove fatal in the pig. I have seen some few cases in which, instead of bronchitis or pneumonia, a severe diarrhea would set in, caused by an irritation of the mucous membrane of the intestines, and it is usually very troublesome and often causes death. I have seen cases of this kind mistaken for hog cholera. Chronic cough is sometimes the result of this disease.

Treatment: In mild cases treatment is not necessary, but the pig should not be exposed to cold or wet, and should have a dry bed to sleep in. In cases where the catarrhal symptoms are severe with fever, which can be known by the appetite being impaired, give epsom salts one to two ounces for a full grown pig, and from a teaspoonful to a dessert spoonful at a dose for a small pig. Boiled flaxseed mixed with the food or given in the form of flaxseed tea mixed with a little brown sugar will be found very useful. If the skin becomes dry and hot give from five to ten grains of nitrate of potassium in the drinking water. If the throat is sore give from ten to thirty drops of the compound syrup of squills three or four times a day. If the pig does not eat and shows signs of weakness, give from one to two tablespoonfuls of good whisky in a little milk three or four times a day; whisky is a very useful medicine in this complaint. Quinine in one to four grain doses in a little whisky and water is also good. In the second stage of the disease if the eyes are very sore they should be bathed several times a day with an emulsion of slippery elm bark. If the pig should be in pain and have diarrhea it should have from five to twenty-five drops each of tincture of opium and spirits of camphor in a little warm milk every two hours until relieved. If bronchial trouble should set in give tar in little balls about the size of the end of one's little finger in the food or a little milk. Five to ten grains of carbonate of ammonia given in cold water several times a day will be found very useful; five to fifteen drops of turpentine is also very good. Careful nursing and stimulants in the form of whisky are usually all that is needed in this disease.

CHAPTER XVIII.

ERYSIPELAS.

This is a constitutional disease characterized by inflammation of the skin with fever.

Causes: Some animals have a predisposition to this disease and it only needs some exciting cause to develop it. It is usually caused by a wound of some kind by which the poison enters the tissue. There is a difference of opinion as to what this poison consists of, but there is no doubt but that it is an organism as it has been found, still the results of experiments are very conflicting. It also makes its appearance without any wound being found on the body and is supposed to be the result of some undue excitement of the skin as from the direct heat of the sun or exposure to severe cold. It has also been ascribed to the kind of food the animal has been fed upon, to a deranged condition of the digestive organs and to rheumatism. Erysipelas has been caused in the human being by stings or bites from insects and the scratch of a pin has produced it, and no doubt but some of the cases of it which we find in pigs are the results of such slight injuries.

Symptoms: There are two forms of this disease, the constitutional and the local. In the first the animal appears dull, refuses its food, the pulse is fast and the breathing hurried. At this stage of the disease it is impossible to diagnose it, it is only when the skin in some part becomes affected that the true nature of the disease is apparent. A pig that has been noticed ailing for a day or two begins to swell on some part of the body, particularly the neck; if the skin is white it will have a red appearance, the redness will disappear on pressure to return immediately the pressure is removed; the swelling rises distinctly above the surrounding parts and continues this elevated march until it ceases, the margin is always abrupt. I saw one case of a valuable sow which had been sick for two days before I was called to see her, the head and neck began to swell and in spite of treatment the swelling extended, keeping an abrupt edge until it spread all over the body and the animal died on the third day. At time of death the body seemed to be twice its natural size. On removing the skin there was great infiltration of serum of a dark color and in some parts black; this extended into the connecting tissue of the muscles and had a very fetid odor, the lungs were found much congested, which was the immediate cause of death, there was considerable effusion in the pleural cavity; the other organs of the body were healthy. In some cases the inflammation rises for three or four days then gradually subsides without any apparent effusion of any kind and terminates in desquamation. In the majority of cases there will be more or less effusion take place, which will exude through the skin, or small vesicles may form and burst, discharging lymph, this is a very favorable sign, or matter may form and cause a large slough. A man told me he had a large pig which swelled in its body and that there was a considerable quantity of matter formed so that the greater part of the skin of one side and part of the belly peeled off and he had the pig destroyed. This was, no doubt, a case of erysipelas. I have seen cases in very fat pigs where the skin of the abdomen loosened from the connective tissue and hung down. There was very little effusion. In time the hair dropped out and the skin contracted into a sort of fold. In time the animal recovered, but it was much disfigured. In cases where the head and face are the parts affected the inflammation often extends to the brain through the nostrils and ethmoidal cells, causing delirium and death.

Treatment: In the early stage a dose of epsom salts is proper and cooling medicine, such as nitrate of potassium in ten grain doses three or four times a day. If there is irritation of the bowels castor oil in one to two ounce doses will be the best cathartic. Should the pulse be fast and full, give a few doses of tincture of aconite three to five drops every two hours until the pulse is reduced both in force and frequency. When there is feebleness from the beginning with restlessness or nerve irritation one grain each of opium and ipecacuanha should be given three or four times a day. Tincture chloride of iron has been found very useful in this disease in doses of from fifteen to twenty drops every two hours throughout the disease without reference to the degree of fever or delirium. The best local treatment is to keep the swollen part constantly wet with acetate of lead lotion, strength half an ounce to the quart of water. The effusion of slippery elm bark has been used with success, but I have had the most advantage from the use of the lead. The animal should be fed on milk and oatmeal and kept comfortable and given all the cold water it will drink. If blisters form they should be opened to prevent them from communication and the consequent loss of the skin, and dress the opened vesicle with acetate of lead lotion. If gangrene should take place in a part it may possibly be arrested by applying a blister over the surface of the part affected. Also support the strength of the animal with beef tea and quinine and whisky. It is seldom that the animal recovers after gangrene sets in in any part of the affected surface.

CHAPTER XIX.

RHEUMATISM.

This is a very common disease among pigs, especially young ones. It is a constitutional disease attended by a peculiar irritation or it might be called an inflammation to which all parts of the body are liable, but it is found most frequently in the hind legs. It is found in two forms, muscular and articular, the former when it affects the tissue of the muscle, and the latter when it affects the structures composing the joints. The nature of rheumatism is not well understood. The profession is divided as to what it really is; some say the offending matter is lactic acid, others that it is uric acid, but none of these substances have been found in sufficient quantities to cause it. Wood says of this disease: "All that we know of the real nature of this disease is that it is peculiar, and that it owes this peculiarity, not to the character of the cause, but to some unexplained condition of the system called the rheumatic predisposition or diathesis. I am inclined to the opinion that this diathesis is in itself a morbid state, in fact, the true disease, and that the irritation and inflammation by which it is recognized are merely symptoms of its full development. That the rheumatic differs essentially from ordinary inflammation is shown chiefly by its shifting character, its disposition to alternate with mere irritation or functional disorder, and the almost entire absence of any tendency to suppuration, even in the most violent cases."

Causes: It is impossible to say what causes rheumatism in the pig, as it is found under all conditions, when the pigs are well kept and when they are not. Cold seems to be an exciting cause of acute rheumatism, moisture increases its effect, thus it is often found in pigs which have to sleep in wet, cold beds, but in the majority of cases something more than cold and moisture is needed. Subacute rheumatism is the form usually found in the pig, although I have seen a few cases of acute rheumatism.

Symptoms: There is lameness of one or more legs which is more or less noticeable according to the severity of the case. The joints often swell at the fetlock. If the swelling is inflammatory there will be decided fever and thus constitute acute rheumatism, causing a rise in temperature, a full, fast pulse, loss of appetite, and an indication that the animal is suffering pain. I have seen some cases of this kind where if the pig was made to move it would squeal with pain. From the effects of the inflammation there may be an increased secretion of the synovial fluid and fluctuation may be noticed in the joints. In some cases the cartilage and tissue of the joints become enlarged and remain so, causing stiffness of the joints. In the muscular form, as well as in the articular, the disease may extend to several muscles or may be limited to one. It very frequently involves several in the same neighborhood and concerned in the same action. There may or may not be inflammation, or only an irritation causing soreness and stiffness with little or no swelling. If inflammation is present there will be pain, swelling and redness, causing high fever. This variety of rheumatism is not confined to the muscles and joints, but may affect any tissue of the body. There is reason to believe that it sometimes attacks the nervous sheaths, producing severe pain along their course or may extend to the nerves themselves, producing spasms of the parts. There is no doubt but that many of the severe complicated nervous disorders, both of external and internal parts, connected with tenderness of the spinal column and the marrow, causing paralysis of the hind legs, are due to subacute rheumatism. In some forms of rheumatism there is a great tendency for it to shift from one joint to another or from one part of the body to another. This form is less likely to cause bad results, as it is only an irritation and not inflammatory. Rheumatism is liable to attack any organ of the body, such as the heart, lungs, pleura, diaphragm, abdomen, stomach, liver, kidneys, bowels, etc. Rheumatism in the acute or subacute form is not a fatal disease unless it affects some of the internal organs, especially the heart, but it is apt to leave complications, especially is this the case in fat pigs. If pigs are unable to rise the constant pressure on the muscles and fatty material from lying on them soon causes them to undergo decomposition and gangrene ending either in sloughing or death. There is a chronic form of rheumatism which affects young pigs which are kept in wet, cold places. This form of rheumatism is most commonly found in the joints, although it may affect the fibrous, synovial or muscular tissue. In this form of the disease the swelling of the joints is not much and to all appearances in some cases not at all. In such cases the muscles often waste away, shrink and become shorter. I have seen young pigs affected with this disease going around in a stiff manner, hump-back, with shrinking of the muscles of the legs, hips and loins; such cases do not grow and are not worth keeping unless they are properly treated and cured.

Treatment: In the early stages of rheumatism give from one to two ounces of sulphate of magnesia or two drops of croton oil in a spoonful of sweet oil; if this does not cause purging in fifteen hours repeat the dose. There is no remedy which will give so much relief in acute rheumatism as a good physic and I have had the best results from the use of croton oil; follow this by giving ten to fifteen drops of the oil of gaultheria in a spoonful of sweet oil or raw linseed oil three times a day. Iodide of potassium in ten grain doses three times a day is also useful, or the bicarbonate of potassium in ten grain doses along with the food three times a day. In cases of inflammatory rheumatism when the pulse is full, fast and strong three to five drops of tincture of aconite every two hours in a little water will be found useful to reduce the fever, then give the oil of gaultheria. In cases of chronic rheumatism I have found arsenic to be of great service, five to eight drops of Fowler's solution of arsenic in the food at a dose three times a day and continued for several weeks.

Local Treatment: When the joints are swollen, hot and tender to the touch use a mixture of one ounce of tincture of opium, one ounce fluid extract of belladonna and half a pint of water, bathe the swollen parts several times a day with a little of this. In cases where the joints are swollen but neither hot nor tender the use of cantharides will sometimes work wonders, strength one part of cantharides to four parts of lard, repeat in a week if necessary. I have seen hogs which could not walk from swelling and deep seated pain in two days after the application of a cantharidine blister be able to walk and soon get well. In milder cases camphorated soap liniment well rubbed in to the swollen parts will often relieve the pain. There are a great many medicines which are used for the treatment of rheumatism, but the above will be found to be the best.

SPRAINS.

Pigs, both young and old, are liable to sprains of the ligaments and tendons of joints which will cause lameness more or less severe and in some cases there will be swelling of the parts. This ailment may be mistaken for rheumatism and if such should be the case no harm would be done, as the treatment recommended for acute rheumatism would be proper. Cases of lameness, although they may be slight, ought to be attended to as any suffering that the pig may be subjected to will cause a reduction of flesh resulting in loss to the owner.

CHAPTER XX.

DISEASES OF THE NERVOUS SYSTEM.

Phrenitis. (Inflammation of the Brain.) Meningitis, Inflammation of the Membrane of the Brain. These two diseases are so much alike that it is impossible in the animal to discriminate accurately between them. There is no doubt but that at times the inflammation may exist separately in either the brain itself or its covering, but it is only on dissection that the true nature of the disease is demonstrated. It is of little practical importance, as the treatment would be the same in either case. In the vast majority of cases both the brain and its membranes are more or less affected at the same time.

Causes: One of the most common causes of inflammation is a rich state of the blood caused by over feeding, short thick necks and a weak circulation, some kinds of food, such as brewers' grains and distillery slops, often produce it. It is often the result of other diseases and injuries, such as kicks and blows, over-exertion, such as being pursued on a hot day. It also sometimes occurs without any apparent cause.

Symptoms: The attack sometimes comes on suddenly or it may be preceded by dulness, loss of appetite and the animal appearing stupid, and as it becomes more pronounced the animal will stagger and have the appearance of being giddy with a wild expression of the eyes; there is often a twitching of the eyelids, sometimes so much so that it may completely close them for a moment, then extend them wide open and so on. At this stage of the disease the pulse is full and hard and the breathing slow; the animal soon becomes very restless and at times will tear substances with its teeth, and sooner or later delirium sets in, the pulse is then full, hard and fast and may be irregular; the breathing is hurried, the skin hot and dry, vomiting is very characteristic of this disease, although I have seen cases where vomiting did not occur. The delirium soon gives way to stupor then to coma, but this is not so common in the pig as in man (exhaustion with convulsion.) We cannot confine the animal in its wild delirious condition and on this account it soon exhausts itself and dies. It is seldom that a pig affected with this complaint will live more than twenty-four to thirty-six hours. I was called to examine a number of pigs which were supposed to be affected with hydrophobia, but which proved to be inflammation of the brain caused by a sudden change of food. The animals had been fed on dry corn in the ear for a length of time without sufficient water and were changed to green corn and corn stalks and in three days after eight of them were affected, I could find no other cause. All the well ones were put back on the dry food and none of them were affected. On the second day after the change from dry food to green corn diarrhea set in and those which had this complaint bad had the brain symptoms. The majority of the affected ones became wild, would run about, some of them squealing, and would attack poultry of any kind and tear them if they got hold of them, they did not attack each other; very soon they would bump up against anything they came in contact with, because they had lost either their sight or sense. I think in most cases both the pigs were in such a state of excitement that it was impossible to get a correct condition of the pulse or respiration; the pulse as found was full and not fast, but irregular; the pupils of the eyes were very much contracted; some champed the jaws and had considerable froth at the mouth as a result, others did not, but all were very restless without a moment of relief; finally they became exhausted or partially paralyzed or a combination of both, first their hind legs, then the fore, they would then struggle on their side or bellies and soon die. I had one of the pigs killed, and made a hasty examination of the head. I found the membrane of the brain much thickened and very vascular, in fact congested; there was considerable fluid beneath the arachnoid membrane, also in the ventricles and the meshes of the pia matter. The brain itself was not much changed, although there was some appearance of congestion on the cortical substance and the cut surface of the medullary portion was thickly dotted with red spots. I found sufficient alterations of the membranes and brain to account for the cause of disease. The lungs and stomach had spots of congestion. All the other organs were healthy. I treated the other seven with a dose of epsom salts and a few doses of tincture of aconite, but only one recovered, the others died during the night.

Treatment of Inflammation of the Brain: In the early stage give a strong dose of epsom salts, two ounces to the adult pig; follow this by giving three to five drops of tincture of aconite if it does not cause vomiting; if it does, it should be discontinued. Ice to the head would be of service, but it is impossible to apply it to the head of the pig. A blister of cantharides to the back of the head may be of use and should be tried.

CHOREA.

Chorea: This disease affects the muscles, causing involuntary contractions in some part of the body. The contractions are not rigid or persistent like those of tetanus, nor quick or jerking like those of convulsions. They resemble somewhat the voluntary movements.

Causes: Defective nutrition has something to do with this disease, as it is often less seen in animals of a feeble condition, some excitable state of the nervous system or the nerves supplying a muscle or a group of muscles. I have seen this complaint affect pigs to all appearance in the best possible condition. It is therefore difficult to ascribe a cause. It seldom attacks very young or very old pigs; from six months to one year is the usual age. The disease does not seem to do any particular harm to the hog as far as its growth and health are concerned.

Symptoms: The first thing noticed is an irregular jerking motion of some of the muscles, especially those of the shoulder, fore leg, or neck. It gives the animal an unsteady gait. When the muscles of the trunk are the ones affected the animal is pulled to one side or the other. I have seen cases where the pig started off its hind parts would drag so much to one side that it would cause the animal to fall over. In slight cases there may only be a jerking of the muscle which is of little inconvenience to the animal. This derangement is easily diagnosed by the absence of fever, coma, delirium and rigid spasms of other nervous diseases which are not present in this one.

Treatment: In the great majority of cases treatment should not be tried. It usually does not affect the health of the animal and it can be fed for market as well as the others. If it should affect the muscles of the mouth so as to prevent the animal from eating as well as is necessary in feeding hogs medicine may be tried. The subcarbonate of iron in half to one dram doses in the food three times a day is a good remedy. The sulphate of zinc in two grain doses increased to three or four if the stomach will bear it or Fowler's solution of arsenic in five drop doses three times a day in food is also good.

TETANUS.

The exciting causes of tetanus are wounds and injuries. A great deal has been written as to the character of the wound, but it is generally admitted that it may take place as a result of any kind of wound, but the punctured wounds, especially if they are in the feet, are more likely to produce it than any other kind. The interval between the reception of the wound and the occurrence of tetanus is very uncertain. According to the statement of some writers it has taken in a few minutes after the operation or wound, but it most frequently takes place as the wound begins to or has almost healed. Idiopathic tetanus is supposed to be caused by something which is likely to disturb the motor nerve system, such as exposure to cold, indigestible substances in the stomach, worms in the intestines and irritation of the urine system. It has made its appearance and the cause could not be determined. It is ascribed to a bacillus tetani which is found in the soil getting into the wound and there developing in the nerve tissue. The fact that tetanus in some cases comes on immediately after a surgical operation would throw doubt on this statement. I intend making a number of experiments on this bacillus tetani this fall and shall publish the results.

Symptoms: The first thing noticed in animals affected with this disease will be a protruding of the membrana nictitans (a membrane of the eye peculiar to animals) when the head is turned to one side. There will be stiffness of the muscles near the seat of the injury; if the animal is made to move, as soon as the muscles of the face become affected it champs its jaws, causing a froth at the mouth; soon the muscles of the back become contracted, drawing the head upwards with the nose poked out, and the ears are pricked upwards and inwards, the back bent downwards and the tail elevated; the muscles are always in a state of contraction, but if the animal is disturbed in any way there will be an extra spasm of the muscles which causes great suffering to the animal. At first the pulse is not much affected, but as the disease advances and the spasms become more severe it becomes faster and harder; the breathing also is faster, especially when a spasm is on, sometimes the spasms are so severe that the animal is thrown down and is unable to rise; the muscles of deglutition are sometimes affected and the animal is unable to swallow.

Treatment: The disease is the result of an over-exhilarated state of the spinal cord and requires a strong dose of medicine which will overcome this condition. I find that bromide of potassium will accomplish this providing the doses are large enough. Give the pig one ounce bromide of potassium at the first dose and half an ounce every two hours until the muscles relax, give the medicine in gruel. Keep the animal hungry and thirsty so that it will suck this into its mouth. Keep the pig in a dark place. There are other medicines which have been recommended, but they are of little use.

PARALYSIS, PARTIAL PARALYSIS.

This is a very common disease in the pig most usually affecting the hind parts. Pathological condition: In most cases paralysis is a mere symptom of a morbid state existing in some other part than the one apparently affected. It may depend upon disease, either in the nervous centers, incapacitating them for the reception of impressions or the origination of influence, or in the conducting filaments which form the communication between all parts of the body and these centers. But it may also be strictly local and depend on an altered state of the terminal nerves. The nerve centers are probably in the gray matter of the brain and spinal marrow and the ganglia. The conducting filaments probably make up the white matter of the brain, spinal cord, and nerves. It follows that the true seat of the disease may be in the encephalon, the spinal marrow, the conducting nerves or the nerve ramifications of the paralyzed part. (Wood.)

I have made a number of post mortem examinations and also examined the spinal cord and have found in some cases the cord and main nerves of the paralyzed parts enlarged and softened with considerable effusion in the sheaths, and in others atrophied and indurated. In some cases I could detect very slight change in the nerve structure. It takes a very slight disturbance in the nerve or its sheath to render it unfit for receiving or sending impressions from the brain or from the nerves in the immediate seat of the disease.

Symptoms: Paralysis may come on suddenly or gradually. Usually the first thing noticeable in the pig will be some stiffness in rising and moving about, with the back somewhat arched, knuckling forward on one or both of the hind legs at the ankles; by degrees this grows worse. If the pig can rise it may be unable to stand, as the hind legs will double under it. In some cases the animal seems to be more or less in pain and if made to move will drag its hind legs. If the animal is not properly treated it gets still weaker until it cannot move and usually dies in from one to two weeks. At first the appetite is not impaired nor the heart's action increased, but as the disease advances the pig will eat but little and the pulse becomes faster and weaker. The only disease which might be mistaken for paralysis of the hind parts is rheumatism. There is no doubt but that severe rheumatism will cause the animal to lose more or less control of its hind legs. In rheumatism the animal will evince more pain on being handled and if excited may even walk for a short distance. There may be a combination of rheumatism and paralysis in cases where the rheumatism affects the sheath of the nerve, but as the treatment of this form of rheumatism would be the same or nearly so as in paralysis there would be no harm done in using it.

Treatment: In the early stages of the disease give the pig one ounce of castor oil and one or two drops of croton oil in it; as soon as the physic operates give eight or ten grains of iodide of potassium three times a day in the drinking water; if the animal will not take it in the drinking water put it in a little water and give it with a spoon. In three days if the animal is not improved give it from two to three drops of fluid extract of nux vomica and from ten to fifteen drops of oil of gaultheria in a tablespoon of sweet oil three times a day. Also rub the back, loins and hips once a day with a little compound soap liniment. Feed the pig on oatmeal mush and milk, with a little of anything else it will take. Also see that the animal is sheltered from the heat of the sun, rain and cold. I have been very successful in the curing of this disease by the above treatment.

HYDROPHOBIA, RABIES, RABIES CANINA.

This is a peculiar disease resulting from the entrance into the system of a poison of a rabid animal; the poison is almost always received by the bite of an animal. The wound usually heals and for some time after no constitutional effects are felt. It is seldom that any symptoms of the disease are shown until about the twentieth day after the bite. Some believe that it may originate in man or dogs independently of any poison entering the system, but it is not likely that it has been real hydrophobia, as there are a number of nervous diseases which somewhat resemble hydrophobia and may have been mistaken for it. There is no doubt but what the saliva contains the germs of the poison, and if it should come in contact with a mucous membrane it will become absorbed.

Youatt mentions that persons have been attacked with hydrophobia in consequence of having wiped their mouths with linen which had been impregnated with the saliva of a mad dog; and a case is mentioned in which the disease originated from an attempt to untie with the teeth a knot in a cord by which one of these diseased animals had been fastened. Horses, oxen, sheep and other animals are said to have contracted the disease by eating the straw upon which mad dogs have lain. We should therefore be very careful if any animal should have an attack of hydrophobia to be on the lookout in handling an animal or anything which it may have come in contact with, in case that any of the poison should touch any abraded surface or mucous membrane. A great many persons who are bitten are never attacked with the disease. It is possible that some systems are not susceptible to the poison. It is said that some animals are more poisonous than others. In man or animals that are bitten by the wolf a much larger portion is attacked than in those bitten by the dog; this is, however, explained by asserted facts that the wolf generally flies at the naked part, as the face and hands, while the dog more often bites through the clothing and in this way the teeth may be wiped off before reaching the skin, and the hair of animals may to a certain extent do the same, but the percentage of deaths of animals bitten by the same dog is much greater than that of man; of one hundred and fourteen cases of persons bitten by mad wolves, collected by Dr. Watson, sixty-seven died; while of fifteen persons bitten by a mad dog only three died. Dr. John Hunter states that he knew an instance in which twenty-one were bitten and only one died. The germ of hydrophobia lodged within the animal body requires time for its reproductive process to be completed, and this process may be hastened or retarded by various conditions which are not easily made out. It usually, I suppose, if conditions are favorable, takes place about the twentieth day, seldom before that period, or it may take from one to three months. Cases have been reported in man in which it made its appearance after several years; in one case ten years.

The symptoms of the disease in the hog resemble very much those in the dog. The animal has at first an altered look, is very restless and frequently changes his position, will often be seen to rub a certain part of its body, and if it should be within reach of its mouth will bite at it, and if the part is examined there will be found a scar, the seat of the bite; it will be somewhat swollen and if the skin was white it will be changed to red, it may even break open again. There is no doubt but what the animal feels it either itchy or painful. The animal has a disposition to pick up and swallow dirt of any kind and occasionally vomit. In some cases it becomes irritable and will bite at anything that comes in its way, or may run at other pigs. There is a flow of saliva from the mouth, caused more or less by a champing of the jaws. The voice of the animal is changed into a spasmodic grunt, somewhat resembling the bark of a dog. There must be some stricture of the throat, as the animal seems to want to get something out of it and will even put its feet up to the corners of the mouth. The breathing is labored and has a peculiar sound, caused, no doubt, from the throat affection. As the disease advances the animal will become more excited, and if at liberty will run around, snapping at everything it meets and sometimes seems as if it were looking for something to attack. At length symptoms of paralysis set in, the legs give way, the lower jaw drops and the animal usually dies about the fourth or fifth day, although I have known them to die sooner.

Treatment is of no use after the disease is once established, and the animal should be destroyed. If an animal is known to be bitten by a rabid animal the parts should be washed and caustic potass applied; this is the best caustic because it will penetrate deeper into the wound than any other caustic. If none is at hand, take a red hot iron and burn the part as deep as possible. If this is well done it will save the animal's life in every case.

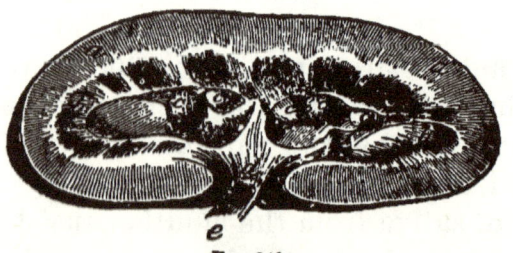

Fig. 142.

Horizontal section of the kidney of a hog. a. Cortical substance; b. Medullary substance; c. Renal papillae; d. Infundibulum; e. Ureter cut across.

CHAPTER XXI.

DISEASES OF THE URINARY ORGANS.

A short description of the kidneys is necessary so that we may have a better idea of the cause and nature of their diseases. The kidneys are two glandular organs situated in the lumbar region of the back. They are composed of a number of tubes and tufts, around which the blood vessels run. The supply of blood to the kidneys is very large for the size of these organs. The tubes begin very small at the surface of the kidney and are very numerous; they soon join one another, becoming larger and finally terminate in a part of the kidney termed the pelvis. At the lower border from this place there is a small duct which leads to the bladder, through which the urine passes. The use of the kidneys is to secrete the water and effete matters in the form of urine and uric acid, which would soon poison the body if it remained there. The blood vessels ramify around the tubes, and the epithelium of the tubes secretes the urine from the blood, and if we consider the quantity of urine which is secreted daily we need not wonder at the size of the blood vessels which go to and from the kidneys. If from want of action, from disease, or otherwise, the kidneys did not secrete this material from the blood the animal would soon suffer from a form of blood poisoning called uremic. The quantity of urine secreted varies very much in the same animal. The weather has a great deal to do with the amount secreted. Animals pass more urine in winter than in summer, as heat increases the quantity removed by the skin and lessens the amount passed by the kidneys. Thick, creamy urine is the result of a sluggish condition of the kidneys, while coffee colored and scanty urine is the result of fever and a partial congestive state of the secretive organs in different parts of the body. An increase in the quantity and a clear or light yellow color denote either over-stimulation or it is seen in cold weather, and on account of it not being irritant the time between voiding it has been prolonged. On the other hand, if it is passed in quantities and often and the animal is very thirsty, it is a symptom of a disease which will be described hereafter. As long as an animal is in good condition and spirits and has a good appetite, no notice need be taken of the urine. Diseases of the urinary organs are very rare in animals, but as they do happen sometimes I shall describe a few of them.

Nephritis (Inflammation of the Kidneys). Causes: Injuries, cold rains, cold water dropping on the loins for a length of time, feeding on brewers' grains and distillery slops, kidney worms, etc. Symptoms: The animal is stiff in its hind parts, pain in the loins if the animal is made to move, or if slight pressure is brought to bear on the loins the animal will squeal. There is loss of appetite and high fever, fast pulse, rapid breathing and elevated temperature. The animal is restless and a few drops of highly colored urine will be passed at short intervals. In a few days, if the animal is not relieved, it will become very weak, staggering on its legs, especially the hind ones; the fever increases, and the brain becomes affected from the blood being poisoned by ureic salts not being eliminated from it, thus causing uremic poisoning. The stomach often becomes affected, causing vomiting, and there will be a strong smell of urine. Treatment: Give a dose of epsom salts, one or two ounces, dissolved in half a pint of cold water. If the stomach is irritable, give three or four grains of calomel and one grain of opium every two hours till three doses are taken. This combination has often the effect of quieting the stomach, so that it will retain salts. When the fever is very high, give from three to five drops of tincture of aconite in a little water every two hours until the fever is reduced. If there is much pain one or two grains of opium should be given. The animal should be encouraged to drink all the cold water possible; barley and ice water may also be given with the bottle. Sometimes sixty drops of tincture of opium, mixed with linseed tea and given as an injection, is very useful. Apply mustard poultices to the loins. When the acute symptoms are passed, give a teaspoonful of spirits of nitrous ether and a teaspoonful of fluid extract of buchu three times a day in a little water. If the heart is irritable, and there is great suppression of urine, bathe the loins with hot water, and saturate a piece of flannel with tincture of digitalis and lay it over them, or give from one to three drops of the fluid extract of digitalis in a little water three or four times a day. When the animal becomes convalescent, give it a teaspoonful of tincture of chloride of iron twice daily in a little syrup. Give the pig anything it will eat.

HEMATURIA (BLOOD WITH THE URINE).

This derangement is sometimes seen in the pig. It appears in two forms, traumatic and idiopathic. Traumatic hematuria is caused by injuries and strains. I have known cases caused by heavy hogs being loaded into cars; also from hogs getting down and being trampled on or squeezed by the others.

Symptoms: The urine is of a blood red color; if there is much hemorrhage it will be of a pink color; very soon after the urine has been passed the blood will separate into clots on the ground or floor; it is therefore easily distinguished from idiopathic hematuria, in which the blood does not separate into clots. In some cases the pig does not want to stand and has some difficulty in rising, and if made to walk will move off stiffly and may show signs of pain. The appetite will be more or less impaired; there is often high fever, fast, weak pulse, and elevated temperature. If inflammation should set in the secretions of urine will be scanty and the bowels are usually constipated.

Treatment: Keep the animal as quiet as possible. Give it one or two ounces epsom salts to clean out the bowels and cool the system; then give ten grains acetate of lead and two grains of opium at a dose, repeat in four hours with half the quantity and so on until the hemorrhage ceases. If the discharge of blood should be abundant give the lead and opium at once. In this case the salts should not be given, as the lead and opium would prevent the salts from physicing. If the fever is high and the pulse full give five drops of tincture of aconite every two hours. Boil flaxseed and make a tea of it and give it cold and as much as the pig will take. If the injury has not been too severe this treatment will cure it.

IDIOPATHIC HEMATURIA.

This form of the disease is observed under a great variety of circumstances. It seems to occur in certain localities and seasons. I have known it to break out among a herd of hogs that had been fed on diseased potatoes, and it has made its appearance among hogs without any apparent cause.

Symptoms: There is a copious discharge of a dark or red colored urine which does not separate into clots on the floor or ground; the animal moves stiffly and is weak in its hind legs, the pulse is fast and rather weak, the breathing is increased in frequency, and in the later stages of the disease is panting or spasmodic, the temperature will be up to 104 to 106, there is loss of appetite, and usually the animal will be thirsty; as the disease advances the pig will be unable to rise from weakness, the pulse will be very fast and weak, the breathing difficult and the animal will either die or be in a state of coma or convulsions. The nature of this disease is not well understood, but it is no doubt caused from faulty nutrition which does not supply sufficient material to keep up a healthy state of the blood, hence a breaking up of the red corpuscles and a passing off of the coloring matter by way of the kidneys. The post mortem examinations have revealed a pale flabby state of the muscles and a lack in the coagulating properties of the blood. From the post mortem appearance the indications for treatment would be a complete change of diet and medicine to stimulate and tone up all the tissues of the body. The food should be fine ground oats of good quality, flaxseed meal and milk, plenty of pure water and a comfortable house. Of the medicines, the tincture chloride of iron in dram doses three times a day will be the best. Also two or three drops of the fluid extract of nux vomica three times a day. As a stimulant take two tablespoonfuls of whisky and dissolve three or four grains of quinine in it, give such a dose three times a day for a few days. Hydrochloric acid in ten-drop doses three times a day is also good. If the bowels are constipated give one or two ounces of castor oil, but it is better to regulate the bowels with the proper kind of food.

ISCHURIA (SUPPRESSION OF URINE).

This is not a common disease among pigs. Suppression or scanty passages of urine is a result of over-stimulation of the kidneys or feeding innutritious food; also such medicines as cantharides, turpentine, wood ashes, etc. In all kinds of fever there will be more or less suppression. When the urine is scanty it is irritant and is passed off in drops. It may also be caused by a plugging up of the kidneys by fat. If there is true suppression of urine, symptoms of blood poisoning will occur, "uraemia." In cases of suppression the animal does not strain as it will do in retention, the urine passing away in drops or in small quantities without any effort of the animal. If this continues for a day or two the animal will show symptoms of being sick, and if not relieved, will soon die, either in a state of coma or convulsions.

Treatment: In such cases find the cause, if possible, and remove it. If this cannot be done (which in the pig is often difficult) treat the symptoms. One of the first things is to give a physic and in this way rid the system of some of the effete matter which is sure to be present in the blood, and nothing is better for this purpose than sulphate of magnesia in doses of from one to two ounces. If there is high fever with a full pulse give three to five drops of tincture of aconite at a dose every four hours. If the pulse is weak and the heart irregular, give from two to three drops of fluid extract of digitalis and a teaspoonful of spirits of nitrous ether in a little water; repeat it every four hours until it takes effect. If there are no symptoms of inflammation, nitrate of potassium, in ten to twenty grains at a dose, put into the drinking water will be found useful. If the pig is stiff and there is not much fever give one dram each of the fluid extract of buchu and spirits of nitrous ether at a dose in a little water three times a day. The pig should have a good supply of cold linseed tea to drink, also milk, with a little lime water in it.

ATROPHY OF THE KIDNEYS.

Atrophy of the kidneys, one or both, has been found in the pig. Usually when one kidney is atrophied the other becomes hypertrophied and it will perform the function of both. Gamgee mentions a case in which one kidney was absent and its fibrous capsule alone remaining distended by a yellow fluid of a strong urinous odor, whereas its fellow was very much enlarged and the animal was in good health.

HYPERTROPHY OF THE KIDNEYS.

This is a very common thing in pigs which have been overfed from an early period. I have examined pigs which have died from other diseases and found one or both kidneys very much enlarged. I have also found such the case in hogs that were killed, which appeared in perfect health and took on flesh very rapidly. Neither atrophy nor hypertrophy can be diagnosed; it is, therefore, when the pig has died from some other disease or has been killed that this derangement is found. Worms in the kidneys have been mentioned already under the head of worms.

RUPTURE OF THE KIDNEY.

I was called to see a fine sow which was very sick. She had the following symptoms: The pig was lying down and it was with difficulty it got up, and when it did so it seemed to be in pain, the appetite completely lost, it would neither eat nor drink, the bowels were regular, it passed a small quantity of urine, which looked natural; the pulse was small and rather weak, the breathing was somewhat faster than normal when she moved, but when she was lying down it was nearly natural. After examining the animal I came to the conclusion that the sow was a little off and that a dose of physic would likely bring her out all right, but to my surprise the owner called to tell me that the sow had died that morning and as I was gathering material for this book I went and made an examination, and was greatly surprised when I found all the other organs of the body healthy except the left kidney, which had been ruptured and surrounded by a large quantity of coagulated blood. In fact, the animal no doubt bled to death, but strange to say that there was not any passed by way of the urine. The only special symptoms to be noted in this case were the small weak pulse and the pain evinced by the animal when it was raised up.

CYSTITIS (INFLAMMATION OF THE BLADDER).

This is a very rare disease in the pig. I have never seen a case of it in the hog, although I have no doubt but that there have been some from injuries, irritating urine, or from inflammation of the surrounding organs. The whole surface may be implicated, or it may only be a part. The neck is the part most commonly involved in the human being.

Symptoms of this disease would be retention of urine, the animal will strain a good deal and the urine will be passed in drops or in small quantities, and the animal will be very restless, as it will be suffering much pain; it will be stiff in its hind parts, there will be fever and swelling of the abdomen, loss of appetite, but the pig may be thirsty, vomiting may occur. The case may get better in a few days. Gangrene may set in, in which case the pain will cease entirely before the close. A case of this kind may assume a chronic form, and the animal may be long affected with a purulent urine or a discharge of pus along with the urine.

Treatment: Hot fomentations would be useful, but it could not be satisfactorily done with the pig. Calomel three grains, opium one grain, made into a pill and given three times a day, and large quantities of mucilaginous, cold drinks would be of great service in such a disease. If the animal is constipated, injections of warm water should be used. If the urine is acid, it should be neutralized by half dram doses of bicarbonate of soda three or four times a day. If it becomes necessary to have recourse to the catheter, an opening will have to be made in the perineum and the urethra slit and a human catheter introduced and the urine removed, then wash the wound with glycerine one ounce, water one ounce, carbolic acid five drops; this is usually all that is required, as the wound will heal readily. There is a chronic form of this disease which I have seen a few cases of. The symptoms of this form of the complaint are a constant desire to pass urine, which is of a whitish color and somewhat turbid, owing to mucous and epithelial scales; in some cases it is very irritating, causing the animal much pain. The quantity of mucus is so great that it nearly blocks up the passage and requires considerable effort on the part of the pig to expel it. If the urine is collected and allowed to stand for a short time the mucus will separate from the urine, thus indicating the nature of the disease. In some cases ulceration takes place in the mucous membrane of the bladder; in this case there will be some hemorrhage, which will give the urine a reddish color. If the disease is allowed to go on the animal loses flesh fast, the general strength gives way, first in the hind legs and then in all. In such a case it is better to destroy the pig.

Treatment: In the early stages of the chronic form give a mild physic, such as one ounce sulphate of magnesia or one or two ounces castor oil. Buchu is very useful, given in the form of the fluid extract in dram doses and sometimes combined with one or two drams of spirits of nitrous ether, diluted with a little water. Thirty drops of turpentine given in some mucilaginous substance three times a day, is useful. Arbutin, in ten to twelve grains at a dose, three times a day, given in a dessertspoonful of glycerine or cod liver oil, is of great service in some cases. The tincture chloride of iron in dram doses three times a day in a little syrup is especially useful on account of its action on the urinary organs. If there is any hemorrhage give from five to ten drops of the oil of erigerontis in syrup three or four times a day. Feed the pig on boiled flaxseed, oatmeal and milk.

VESICAL RETENTION.

Retention of urine in the bladder is caused by some obstruction to the passage or inability of the walls of the bladder to contract on its contents. Obstruction may arise from inflammation of the mucous membrane at the entrance of the urethra. It may arise from spasms of the neck of the bladder. Other causes are tumors, stones, accumulations of mucus and stricture of the urethra.

Symptoms: The animal is making constant ineffectual efforts to pass urine and is in great distress. On examination of the back part of the abdomen, just in front of the pubis, a tumor will be felt, forming a somewhat round, well-defined tumor, and it is sometimes visible. In very fat pigs it may be difficult to feel in this locality; in such a case the dulness upon percussion over the region which it occupies, contrasted with the resonance of the surrounding space, will be sufficient to diagnose it. In this affection the pig becomes feverish and restless, until at length a portion of the urethra or bladder gives way and the urine escapes into the peritoneum with fatal results.

Treatment: It is not very easy to find the cause in the pig. Our first efforts will be to try to relieve the bladder of some of its contents by gentle pressure and the application of hot water to the back part of the abdomen and the perineum. If this should fail, then an opening must be made into the urethra by cutting through the perineum and a human catheter passed and the water drawn off. This will give the animal relief. Then find the cause, if possible. If from calculi or coagula in the urethra, remove them. If from inflammation of the neck, give a dose of epsom salts and bathe with warm water. When spasms of the neck of the bladder are the cause of retention, give injection of warm water with a little opium in it, one dram of opium to the ounce of water; repeat this every two hours; apply chloroform or ether to the nostrils until the animal is slightly under the influence of it. If the retention should arise from want of power in the muscular coat, give stimulants and tonics, such as two drops of fluid extract of nux vomica and thirty to sixty drops of tincture of iron at a dose, in a little syrup, three times a day.

INCONTINENCE OF URINE. (ENURESIS.)

In this derangement the animal has lost the power of controlling the sphincter of the neck of the bladder, and the urine passes away involuntarily. In some cases the bladder may be inflamed or greatly irritated, and the presence of even healthy urine could not be tolerated and would be passed off as soon as it reached the bladder. When it is caused from loss of power the animal will not be feverish or be suffering any pain, but should it be caused by inflammation or irritation there will be more or less fever and pain.

Treatment: If from weakness and loss of power give two or three drops of fluid extract of nux vomica and from thirty to sixty drops of tincture chloride of iron at a dose in a little syrup three times a day. Ten to fifteen drops of turpentine at a dose in oil or syrup is often useful. Five drops tincture of cantharides at a dose in a little water three times a day has often good results. If it is caused by inflammation or irritation remove the cause, if possible, and give medicine required to reduce the inflammation, such as one to two ounces epsom salts and two to three drops of fluid extract of belladonna three or four times a day, or after the physic has operated give one to two grains of opium and three to four grains of calomel three times a day, and encourage the animal to drink flaxseed tea, effusion of slippery elm or barley water. Feed on an oatmeal and milk diet.

URINARY CALCULI. (GRAVEL.)

This affection is very seldom met with in the pig on account of the great majority of hogs being sent to market before or by the time they are one year old. These deposits may form in the kidneys and pass along the ureters to the bladder and out through the urethra without causing any great inconvenience, but should they stick and block up the tubes they would cause very great trouble. The symptoms would be irritation and retention of urine; small stones may remain for a long time in the bladder without causing any very great disturbance; if such should be suspected the animal should be fed for the market. I have met with a few cases of prepucial calculi in the castrated hog, caused by the urine running over the anterior part of the prepuce, leaving a deposit of lithic acid gravel, causing irritation and swelling of the prepuce (sheath), causing considerable disturbance to the animal, such as loss of flesh and some stiffness. The swelling of the prepuce and some stiffness attracts the attention of the owner and an examination of the parts will reveal the nature of the trouble.

Treatment: Cast the hog on its side and remove all the deposit with the fingers or a pair of forceps. When this is accomplished it will be found that the lining of the sheath is red and sore. After it has been well washed and dried, mix one ounce glycerine, one ounce water and thirty grains of tannic acid; a little of this should be applied once a day until the soreness and swelling have disappeared. After it is cleaned out and dressed the animal will be much relieved.

CHAPTER XXII.

WOUNDS.

These are of frequent occurrence in the pig, and if they are severe they are difficult to heal, as it is impossible to keep the animal from rubbing the sore, thus irritating it. Bandages cannot be applied with any satisfaction, as the animal will tear or bite them off. If the wound is caused by a sharp instrument and is cut lengthwise on the muscle it will be advantageous to sew it up. First clean the wound of all foreign substances, such as dirt, hair, or pieces of wood. If there is much hemorrhage it should be stopped before stitching by applying cold water to it. If the vessel is large it will require to be taken up and a ligature put on, then either use pins, if the wound is small, or catgut or silk thread if it is large, and pour over it a little of the following lotion: Acetate of lead half an ounce, carbolic acid half an ounce, water one quart. If the wound is ragged and torn or cut crosswise on the muscle, there will be no advantage from stitching it, as the ragged portions have to slough and thus open the wound. If the muscle is cut across, its fibres will move every time that the animal moves the muscle, and thus prevent union of the cut surface.

Treatment: Great care requires to be taken not to irritate a fresh wound, either by washing or using strong medicines, as they will prevent the healing process and cause inflammation. Wash the wound by allowing tepid water to run over the injured surface until it is clean, then use acetate of lead half an ounce, carbolic acid half an ounce, water one quart; apply a little of this twice a day. If the wound should become unhealthy mix one dram chloride of zinc in half a pint of water and apply a little twice a day. For wounds which are superficial, such as a piece of skin pulled off, mix one ounce oxide of zinc with two ounces vaseline and apply a little once a day, or mix one ounce vaseline, one ounce water and three grains tannic acid and apply a little twice a day with a feather. For proud flesh use terchloride of antimony, and after it has been destroyed use any of the above lotions. For slight wounds in summer the application of a little tar will keep the flies off and also dirt, and assist the healing process. Such strong medicines as turpentine should not be used on a fresh wound. To destroy maggots mix half an ounce of carbolic acid in one ounce glycerine and apply with a feather. Deep or punctured wounds should be examined with a probe to find the depth and direction, and at the same time to feel if there is any foreign body in them. Punctured wounds are the most dangerous of all to prevent inflammation and mortification. I have known cases of this kind to cause death, which a little liniment would have prevented

Treatment of Punctured Wounds: Mix carbolic acid two drams, water six ounces, dip a strip of soft muslin and press it into the wound with a probe, then draw it out and put in a fresh one; let this stop in for a few hours, then draw it out and put in a fresh one; do this at least three times a day until matter forms, then all danger is past. The reason I use the strip of muslin dipped in this strong carbolic solution is so that it will be sure to reach the bottom of the wound, acting as a disinfectant, preventing inflammation and blood poisoning. When matter forms, clean it out with a syringe and inject a little of the above lotion once a day.

INJURIES.

Broken bone in the pig will heal as fast as in any other animal, but it is impossible to confine the animal for treatment. In case of simple fracture it will usually get well itself, but where the bone is splintered and the flesh lacerated, it is best to destroy the animal. The injury sometimes only bruises the flesh, and it will right itself. In other cases there will be effusion of serum under the skin, causing a considerable soft, puffy swelling. This will have to be opened and the fluid pressed out and a little of the chloride of zinc lotion injected once a day for a few days. If this is not done the sack will fill up again.

The part may be bruised to such an extent that an abscess may form; if so, as soon as it becomes soft open it in the most prominent part, and after the pus has escaped wash it out with warm water and inject a little of the following twice a day: Acetate of lead half an ounce, carbolic acid two drams, water one pint.

Ulcers and fungous growths are found on the body of the pig. The ulcers should be treated by applying nitrate of silver to them to destroy the unhealthy surface, then use any of the healing lotions mentioned for wounds. Fungous growths appear about the legs and abdomen and at times on other parts of the body. They are usually very foul smelling and red or of a dark color, usually covered with a dark colored scab. They bleed very easily and therefore sometimes receive the name of "bleeding fungus."

Treatment: If there is any neck to the fungus tie a small cord tight around it, and if it does not slough off in a week tie on another; when it does slough off, apply a little terchloride of antimony to it to destroy the roots, then apply a little of the following: Oxide of zinc one ounce, vaseline two ounces. If a cord cannot be tied on scrape off the scab and apply a little terchloride of antimony to it every second day until it is lower than the surrounding parts, then use a little of the oxide of zinc ointment. All unhealthy sores should be dressed once or twice with the antimony. This will bring a healthy action to the part.

CHAPTER XXIII.

DISEASES OF THE GENERATIVE ORGANS.

The pig is not subject to diseases of these organs for the same reasons that have been mentioned before, that they are brought to maturity and the greater number of them sold by the time they are two years old or less.

Difficult Parturition.—This is sometimes met with in the sow, and on account of the passage being too small to admit the hand, in the majority of sows, it is sometimes with great difficulty that we can relieve them. If a sow has been straining for some time and making no progress, it will be necessary to examine it to find the cause. Causes: The parts may not be properly dilated, or the passage may be too small to allow the foetus to pass, or the parts may not be strong enough to expel the foetus. In other cases several may be expelled and one or more seem to block up the passage. Every farmer and stock breeder should have a pair of parturition forceps for the pig. They can be obtained from Frank Wagner & Co., manufacturers of revolving and obstetrical instruments, Mansfield, Ohio. These instruments are the best in the market at the present time.

Figure 3. Insert and then open a little, any of our string instruments and you have a repeller that will never slip nor will it mutilate as do the metal repellers now in use.

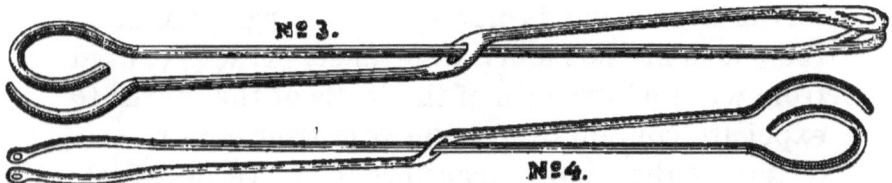

Forceps No. 3 and 4, length about 20 inches. These forceps are designed for pigs and lambs, in many places where malposition has taken place, both numbers are needed, but no case has ever been found so difficult that these forceps would not effect speedy deliverance. They are entirely and easily handled from the outside.

The forceps are useful for two purposes: To dilate the parts by smearing it with fluid extract of belladonna, then introducing it into the passage, and by opening and closing the forceps, and the belladonna acting on the parts, dilation sometimes takes place rapidly. When this is accomplished there will be no further trouble. If it is caused by the foetus being too large, warm the forceps and rub on a little lard, and introduce them and get a hold of the foetus, and by gentle traction you will be able to remove it. I have removed a number in this way and they lived to grow into fine pigs. The foetus of the pig is not like the larger animals; their legs are not much hindrance. If it is caused from want of strength of the walls of the uterus to expel its contents, give the sow from one to two drams of the fluid extract ergot of rye in a little water every half hour until you have the desired result. In cases where the forceps have been used it is always safe to wash out the parts with tincture of opium one drachm, carbolic acid thirty drops, water one pint. Inject warm water first to clean it out, then inject the lotion. This will act as a soothing antiseptic and will greatly soothe the irritated parts. If there should be any laceration of the parts they should be dressed with a little of the above lotion once or twice a day.

INVERSION OF THE UTERUS.

This sometimes takes place in the sow, and when it does the parts should be well washed and all particles of dirt removed, then bathe it for ten

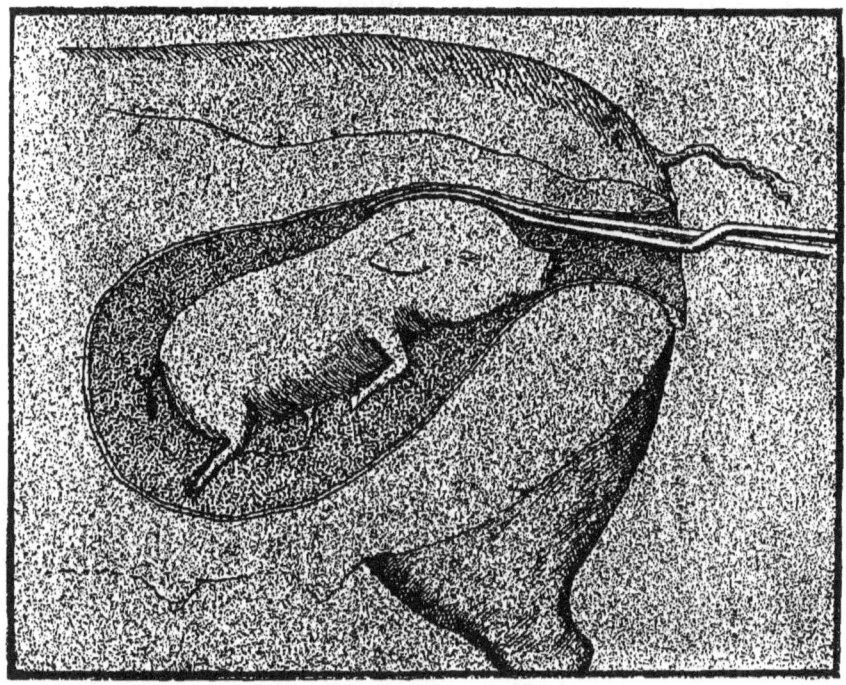

Plate No. 4.

This diagram shows the manner of inserting our blade forceps No. 3. All other forceps must be inserted open with one blade at each side of the head of the young, a thing often very difficult if not impossible when the head is wedged in the pelvis. There is, however, always room at the top of the head, where our long, neat and perfectly shaped blades may be easily inserted. The blades are then revolved to the sides of the head, when owing to the joint, each blade adjusts itself flatly on the head, allowing the severest pull without injury to the young.

minutes with acetate of lead two drams, tincture of opium two drams, water one pint. Then take a piece of soft muslin and fold it into several thicknesses and put it over the ends of your four fingers, which should be made into the form of a cone, and by gentle pressure on its center it can be pressed into its place. Remember it is inverted and the pressure must be on its center so that it will turn in. After this has been accomplished take a strong pin and pass it through from side to side of the passage, then tie a piece of string in the form of the figure eight; that will keep the part closed and prevent the uterus from being pressed out again. The pin is to be left in for a few days, when it can be drawn out, leaving no sore. If the sow is very restless she should be given a dose or two of opium, two grains at a dose. Feed her on oatmeal and milk food for a few days. If the sow does not do well after such cases, give her twenty drops tincture chloride of iron, and a teaspoonful each of tincture of ginger and gentian at a dose, in a little syrup two or three times a day for a week.

MAMMITIS. (INFLAMMATION OF THE UDDER.)

This is not a common disease in the sow, but it does take place at times.

Causes: Injuries to the udder; also, the over-accumulation of milk, and chills.

Symptoms: The udder is swollen and hard. If the skin is white it will be red and very tender to the touch. When this takes place in the sow it is

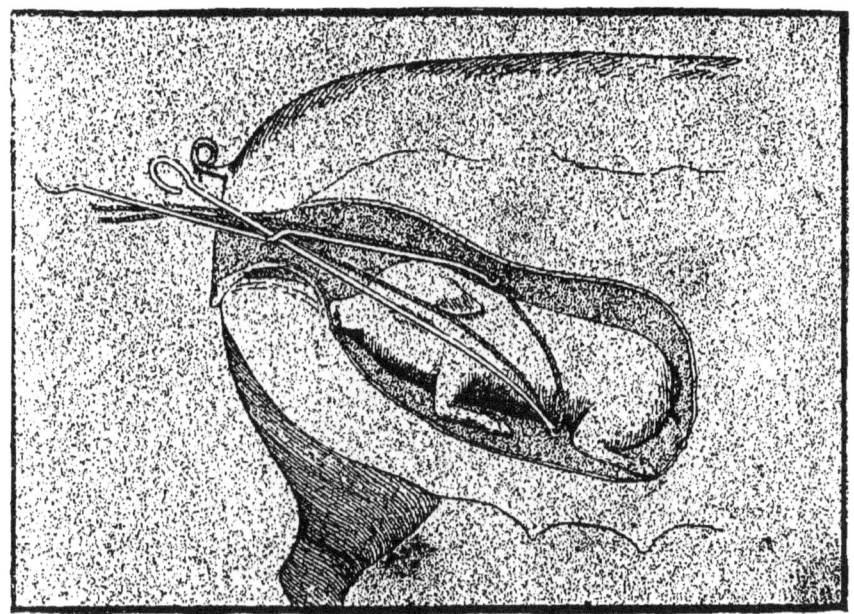

Plate No. 5.

Shows the manner of adjusting our invaluable string carrying forceps No. 4. In numerous cases the pig is found settled so low behind the pelvis that old hinge forceps will not touch it. In such cases this is the only instrument that will reach the pig, and this only in the following way: One arm is held stationary while the other long and properly curved arm is run far back and down until it can be brought under the pig, drawing the cord around in a loop, when the pig can be raised.

usually sick, with considerable fever, loss of appetite and constipated bowels.

Treatment: Draw off as much milk as possible, although this is no very easy matter in the sow; the inflammation causes the milk to coagulate and it separates into curd and whey. If she has young pigs allow them to suck; if not, draw off as much as possible with the fingers. Foment the parts well with hot water, then use acetate of lead half an ounce, tincture of arnica two ounces, water one quart. Bathe three times daily with hot water and apply the lotion after each bathing. Give from one to two ounces of epsom salts; follow this by giving ten grains of nitrate of potassium in a little water three times a day. If matter should form, which is known by the part becoming soft and pitting with the fingers, either open it with the knife or let it break of its own accord. When it breaks inject water into it to clean it out, then inject carbolic acid one dram, water half a pint. Clean the parts out twice a day with a little of this. If the parts become hard or caked, rub on iodine one dram, vaseline one ounce. Do this twice a week until the part becomes soft or the hardness disappears; also give from eight to ten grains of iodide of potassium in a little water twice a day for a week. If any of it should become mortified, which can be easily seen by its bluish or black appearance, it will have to be removed with the knife and the part dressed with a little peroxide of hydrogen, one part to four of soft water, twice a day; also, give twenty to

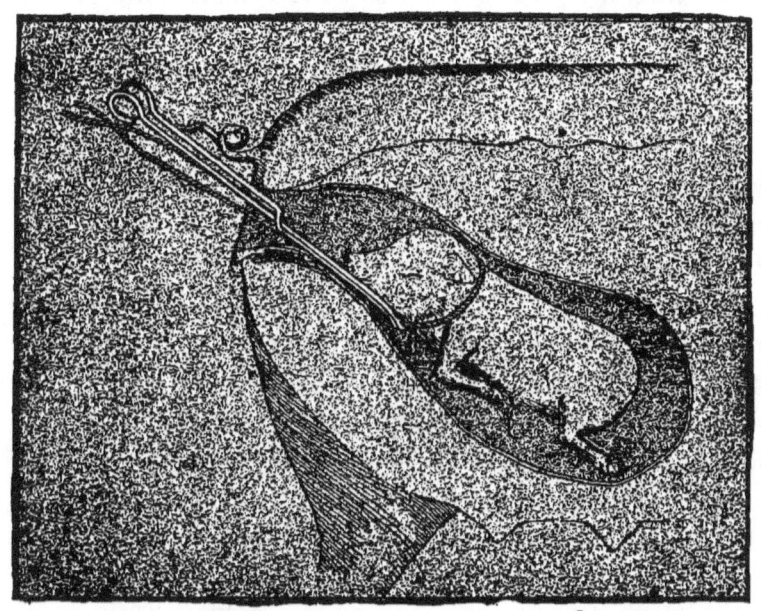

Plate No. 6.

This cut shows how our No. 4 can be used in common cases where the head is up, in place of the blade forceps, No. 3.

thirty drops of tincture chloride of iron in a spoonful of syrup three times a day.

SORE TEATS.

Occasionally the teats of the sow become inflamed and are red, swollen and sore, and she does not want the young ones to touch her. This usually makes matters worse, as the udder at the base of the teats becomes involved on account of the accumulation of milk. This trouble is caused by the teats coming in contact with dirt mixed with urine, which irritates the skin, causing it to crack and inflame. Mud and water are not likely to do this. It is also caused by the sow traveling through poisonous weeds when they are wet with dew or rain. To avoid this trouble keep the sty of the sow clean, and if possible, have all obnoxious weeds removed from the pasture where nursing sows are kept.

Treatment: Bathe the parts well three times a day with acetate of lead half an ounce, water one quart, then rub over the affected parts after bathing, glycerine two ounces, tannic acid twenty grains, water four ounces; shake up well before using. If the animal is feverish, give her one to two ounces epsom salts, dissolved in half a pint of water; also give ten grains of nitrate of potassium in the food or drinking water two or three times a day. This derangement is sometimes caused by the young pigs having sore mouths. If such is the case, sponge the mouth three times a day with a lotion made by putting a teaspoonful of boric acid in a teacupful of soft water.

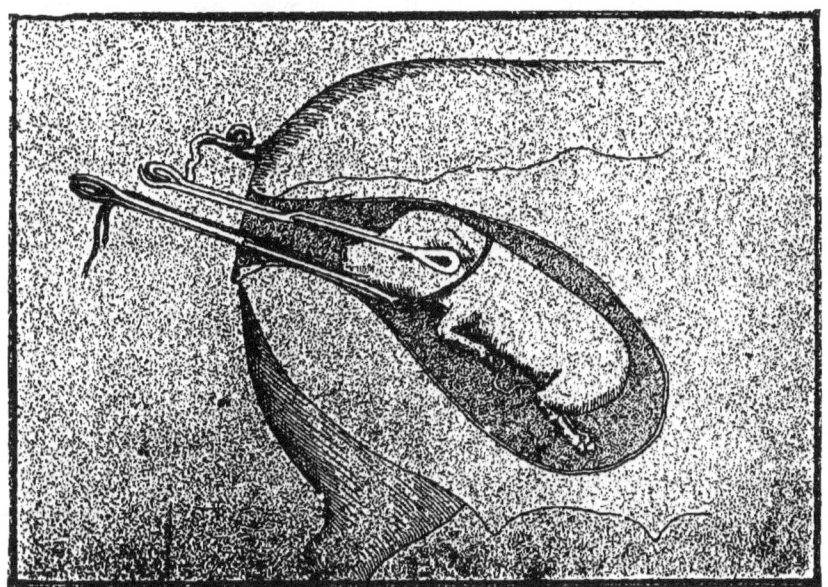

Plate No. 7.

The above cut shows how to manage very difficult cases, in which the pig is so large that no other instruments, and perhaps neither one of our own, will give deliverance. In such cases we have never failed when using both forceps, the string carrier to draw and the blades to work and lessen the head.

ORCHITIS (INFLAMMATION OF THE TESTICLES).

This disease is caused by injuries to the part, such as blows, bites from other pigs, wounds penetrating the testicle. It is also caused by feeding too much stimulating food in hot weather, or medicines that stimulate the generative system, such as cantharides, rue, tansy, demeana.

Symptoms: Swelling of the part, which is hot and tender to the touch, the swelling often extending down the legs; the animal suffers pain when made to move. This disease can be easily distinguished from hydrocele (water in the pouch) by the swelling being hard and hot and very tender, while in hydrocele it is soft and will fluctuate under the fingers and is not painful to the touch.

Treatment: Give the pig two ounces epsom salts; follow this by giving ten grains of nitrate of potassium three times a day in a little water. If the appetite is good, put it in its food. Bathe the part three or four times daily with hot water and after each bathing use a little of the following: Acetate of lead half an ounce, tincture of arnica two ounces, water one quart. If matter should form, which can be known by the part becoming soft and pitting with the finger, then open it with the knife, squeeze out the matter and wash it out with water, then inject a little of the following: Peroxide of hydrogen one ounce, water four ounces. Do this twice a day to heal it. If it will not heal it will be necessary to castrate the animal.

HYDROCELE (WATER IN THE SCROTUM.)

This derangement is sometimes the result of inflammation of the scrotum or by the walls of the scrotum being bruised.

Symptoms: The scrotum is swollen and will fluctuate under the fingers, and the testicles can be felt floating in the water.

Treatment: If it is not interfering with the animal's health and not increasing in size it will be better left alone. Astringent medicines may be tried, such as tannic acid, twenty grains to the ounce of water. The proper treatment is to draw off the fluid with a hypodermic syringe, the nozzle of which is passed through a solution of carbolic acid. But this should be done only by a veterinarian.

STERILITY.

This derangement happens more commonly in the well bred animal, especially in those that are "forced" by overstimulating food. The animal gets into such a plethoric state that it sometimes blocks up the fine tubes connected with the organs of generation. There are several other distinct causes: Disordered ovaries, obstructions to the fallopian tubes, a morbid condition of the uterus, hardening of the neck of the uterus. In the pig, on account of not being able to make an examination by the hand, it is very difficult to find the cause, and if we did it would likely be impossible to remove it. Sows should be kept in good growing condition, but avoid having them overfat. If the animal takes on fat very easily and will not breed give her two ounces epsom salts dissolved in half a pint of cold water at one dose; follow this by giving ten grains of iodide of potassium twice a day in her food for two weeks. By this treatment we may succeed in absorbing materials which have blocked up some of the tubes. If the animal is weak and in poor condition give good food and twenty to forty drops of the tincture chloride of iron twice a day in the food.

CHAPTER XXIV.

HOG CHOLERA AND SWINE PLAGUE.

Hog cholera and swine plague are both very fatal diseases, destroying great numbers of hogs yearly, especially in the corn-growing States. It attacks pigs at all ages, but shoats seem to be more liable to it than older ones; the older ones have more power of resisting the virus than the younger ones. There is no doubt but that the disease is the result of a bacteria, but why this bacteria should make its appearance is not easily understood. In the several outbreaks which I have studied it has acted very peculiarly; as an illustration—it first made its appearance on the farm of a Mr. A., destroying nearly all his hogs, young and old; Mr. A.'s pigs were fine bred and well kept in the way of cleanliness, pure water, good pasture, food, principally corn in the ear; the pigs were all fat. Their neighbors on all sides had herds of hogs, some well bred, others not; some were allowed to wallow in stagnant pools, others kept clean; none of these took the disease. Messrs. B., C. and D., living some three miles distant, lost very heavily from this disease. I have known cases where a man kept only two or three hogs on his place and the disease would carry them all off. It is generally admitted that large herds of animals kept together are more liable to disease than when only a few are kept in the same place, and I think there is some truth in it, but it does not hold good in hog cholera, as it will make its appearance in all sorts and conditions of hogs (the "land pike" excepted). The reason, no doubt, why it does not usually kill all in a herd is that some have more resisting power or that they in some way become immune to the action of the bacteria. Swine plague is just as fatal a disease as hog cholera and both may be present in the same outbreak. The symptoms of the disease are nearly the same and it is only by the use of the microscope that the difference can be ascertained (figs. 18 and 19); but it is of little importance to the swine grower whether it is hog cholera or swine plague, as the management of both diseases are alike. It is said that hogs which have resisted an attack are immune from future attacks; this may be so in some cases, but not in all, as I have known hogs in a herd of swine attacked with hog cholera, a few of which escaped but were attacked the following year and died. Another peculiarity of hog cholera and swine plague is that some years it is much more virulent than others, sometimes destroying ninety to one hundred per cent; at others it may not amount to more than twenty to thirty per cent. The first of the outbreak is always the most severe; towards the end the majority attacked recover.

Symptoms: The sudden death of one or more hogs calls the attention of the swine grower to the fact that something serious is the matter with his pigs (alhough other diseases may have caused this). In the early stages of the disease the pig is noticed to be dull, will neither seek food nor water; it likes to hide itself, lies down most of the time, its head is low and the ears will be lopped; often the signs of pain will be well marked by the constant movements of some parts of its body, or the first symptoms may be cough with a little discharge from the eyes and nose, the exudation from the lids of the eyes is of a gummy nature, which sometimes glues the edges of the lids togther so firmly that the animal cannot open them; at this stage of the disease the appetite may not be in the least impaired. I have made post mortem examinations in this form of the disease and, notwithstanding the animal having a good appetite, I have found well marked ulcer tufts in the large intestine. It lies mostly on its breast and abdomen and may remain in the position for hours if not disturbed. In some cases there will be violent vomiting and the brain becomes affected and the animal may become frantic, or it may lie in an unconscious state until it dies. In the early stages of the disease the feces are normal, but very soon a very foetid, black or dark diarrhea sets in. The pulse rises to one hundred to one hundred and twenty-five per minute, and the heart beats are barely perceptible. There is a peculiar spasmodic breathing in all cases where the lungs become congested.

There is a marked weakness of the hind legs; the animal staggers, its legs crossing each other, but differs from paralysis in its being able to move them until the last. Some time before death there are patches on the skin of a bluish or purplish color, especially on the inside of hind legs. If these patches are pressed they will become pale, which does not occur in other diseases where the skin becomes discolored, such as in erysipelas. The temperature of the body is at first increased, but soon falls below normal, and I have seen in a few cases, dark blood oozing through the skin. In a number of cases the animal dies in from three to six hours, others live for several days. Although an animal may die in from three to six hours from the time it is first noticed to be sick, there is no doubt but what the animal has been ailing more or less for several days before it is actually taken down sick. I have made post-mortem examinations of pigs, which, to all appearance, seemed well; but on opening them, all the characteristics of the disease were present in an undeveloped form. It is ncessary to make a post-mortem of the first hog that dies to enable us to form a correct opinion as to the nature of the disease.

Post-mortem appearance: On removing the skin there is usually found an accumulation of serum often mixed with blood, causing red or black spots; this is the result of the plugging and rupture of small blood vessels. I have seen some cases which resembled that which is produced when a hog has been roughly handled on being shipped. In most cases of hog cholera all the organs of the body are more or less red spotted, caused by hemorrhage of a greater or less extent. The extravasation of blood is found most abundant in the lymphatic glands and the serous membrane of the chest and abdomen. The cases vary very much, sometimes the intestines, both outside and inside, the surface of the lungs, liver, heart and kidneys will be covered with an exudation of blood. On the internal surface of the large intestines in nearly all cases of hog cholera and swine plague there will be found a number of tufts which receive the name of ulcers; they are elevations of a dirty gray or sometimes a yellowish gray; they are more or less hard or tough when cut with the knife; their surface is tufty, somewhat like the top of a wart on the human hand after it has been soaked in water for a while; this surface is covered with a yellowish substance, which is easily removed by scraping off with a knife. These growths extend in some cases through the intestine. In most cases the lining of the intestine and its walls are black in the vicinity of the growths. These tufts may be single or a number of them may be attached to each other, covering a surface of from one to three inches in length; they are usually found in the cecum, upper half of the colon and on the ileo-cecal valve. Very often the small intestine is found more or less inflamed and the glands enlarged; in some cases the spleen is enlarged; the lungs are usually more or less implicated, especially is this the case in swine plague. I have found some cases of swine plague in which there was pleurisy with hydrothorax to a considerable extent, also considerable effusion of fluid in the abdominal cavity. There are often indications of heart derangement, such as effusions of fluid and blood clots, and in chronic cases enlargement of the walls of the heart. In swine plague the liver is often found in a very deranged condition, of a bluish gray color, soft and falling to pieces when handled, especially in chronic cases. Hog cholera has usually more intestinal lesions, and swine plague more lung and liver affections. The hog cholera germs are very vigorous and more hardy than those of swine plague. They are capable of multiplying and living for a long time in water, ponds and streams; they may live in the earth and rubbish for three months or more. Swine plague germs, on the other hand, are much more delicate and easily destroyed. In order that they will multiply and grow the temperature must be more constant and the surrounding media more favorable than is required for the germs of hog cholera. It is said that the swine plague germs are widely distributed in nature and probably present in all herds of swine, but they are not deadly to these animals except when their virulence has been increased or the resistance of the animals diminished by some unusual conditions. The hog cholera germs, on the contrary, are not usually present and must be introduced from infected herds before the disease can be developed. This may or may not be true, as it is impossible to find out the cause of the beginning or end of a contagious disease. If it were possible, in all likelihood, we would have the key to prevent the outbreaks of the disease, which we certainly have not at the present time. Hog

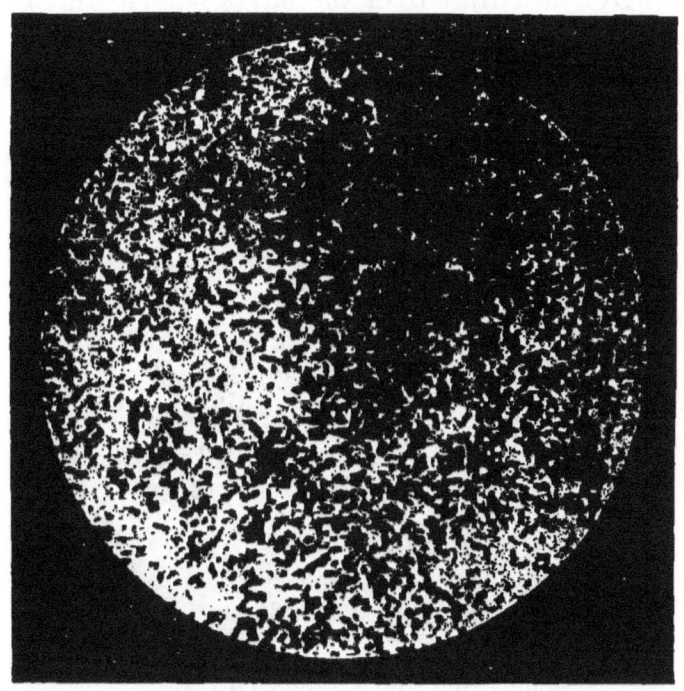

Figure 18. Hog Cholera.

Photograph by Dr. T. J. Burrill, with Zeiss apochromatic 2 mm. objective and number 8 ocular; X1000.

cholera germs are slightly larger and more elongated than those of swine plague; they are provided with flagella or long thread-like appendages, which enable them to move rapidly in liquids; while the swine plague germs (Fig. 19) have no such organs and are unable to move except as they are carried by the liquid in which they float.

228 DISEASES OF THE HOG.

Sanitary Measures: When hog cholera breaks out in a neighborhood it will be wise not to frequent the place in case of carrying the disease, but to scatter new lime over the parts where your hogs frequent most and to sprinkle their sleeping places with a solution of crude carbolic acid, strength one

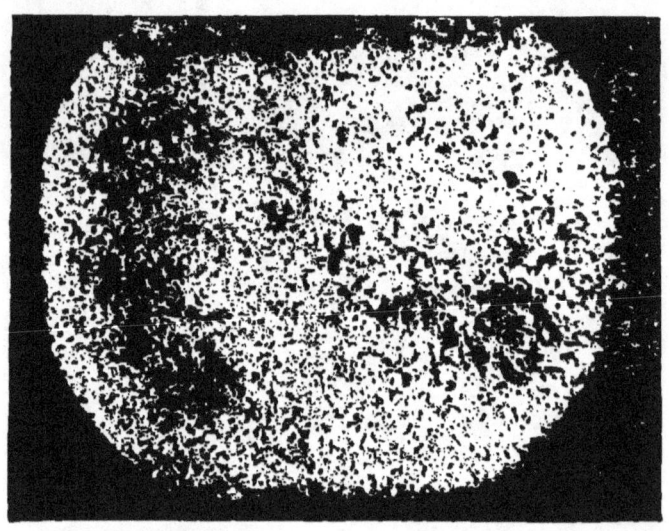

Figure 19. Swine Plague.

Photograph by Dr. T. J. Burrill, with Zeiss apochromatic 2mm. objective and number 8 ocular; X800.

ounce to the quart of water. Care should be taken in the purchase of hogs to see that they have good, healthy constitutions and that hog cholera has not prevailed for some time in the neighborhood where the pigs were raised. I think the best prevention is to have strong, healthy hogs. Read the introduction to be found on the first page of this book. All pigs which die of cholera or swine plague should be buried deep and the body covered with slacked lime before being covered with the earth, and the part where the sick animal was before and where it died sprinkled with the carbolic lotion above mentioned. When the disease is in the neighborhood and before it makes its appearance or after it has done so, each pig should have ten drops of strong nitro-hydrochloric acid in its food twice a day for a week. It is wise when the disease breaks out on the farm to separate all the well ones some distance from the affected ones and put them under the acid treatment. I have seen excellent results from this.

Treatment: A great many hogs can be cured of this disease if only properly treated, and there is no doubt that in many cases it would not pay to undertake it; but when a man has a few or a number of valuable hogs it would pay to have them treated. The pigs should be put into a comfortable well ventilated place free from draughts, and in the early stage each should get two ounces of castor oil; this should be followed by giving ten drops of nitro-hydrochloric acid well diluted with water three times a day; also give half an ounce of spirits of turpentine in a little sweet oil once a day; continue this for three or four days, then give four grains of quinine dissolved in two tablespoonfuls of good whisky, with the same quantity of cold water at a dose, three times a day for four or five days if necessary. This treatment I followed up last year with excellent results. Some very valuable sows, which were not able to rise, recovered by the use of the quinine and whisky. As soon as the animal is able to eat feed on milk and eggs, boiled flaxseed, oatmeal, etc. The nitro-hydrochloric acid is one of the best blood purifiers that we possess; the turpentine acts as a local disinfectant and stimulant as well as a healer to the intestine; the quinine and whisky are tonic and stimulant, and may in some way stop the development of the bacteria. To be successful it must be well done.

www.ingramcontent.com/pod-product-compliance
Lightning Source LLC
Chambersburg PA
CBHW021828230426
43669CB00008B/904